AF566767

ELLIS HORWOOD SERIES IN APPLIED SCIENCE AND INDUSTRIAL TECHNOLOGY

Series Editor: Dr D. H. SHARP, OBE, former General Secretary, Society of Chemical Industry; formerly General Secretary, Institution of Chemical Engineers; and former Technical Director, Confederation of British Industry.

This collection of books is designed to meet the needs of technologists already working in the fields to be covered, and for those new to the industries concerned. The series comprises valuable works of reference for scientists and engineers in many fields, with special usefulness to technologists and entrepreneurs in developing countries.

Students of chemical engineering, industrial and applied chemistry, and related fields, will also find these books of great use, with their emphasis on the practical technology as well as theory. The authors are highly qualified chemical engineers and industrial chemists with extensive experience, who write with the authority gained from their years in industry.

Published and in active publication

PRACTICAL USES OF DIAMOND
A. BAKOŃ, Research Centre of Geological Techniques, Warsaw, and A. SZYMAŃSKI, Electronic Materials Research and Production Centre CeMat'70 S. A., Warsaw
NATURAL GLASSES
V. BOUSKA *et al.*, Czechoslovak Society for Mineralogy & Geology, Czechoslovakia
POTTERY SCIENCE: Materials, Processes and Products
A. DINSDALE, lately Director of Research, British Ceramic Research Association
MATCHMAKING: Science, Technology and Manufacture
C. A. FINCH, Managing Director, Pentafin Associates, Chemical, Technical and Media Consultants, Stoke Mandeville, and S. RAMACHANDRAN, Senior Consultant, United Nations Industrial Development Organisation for the Match Industry
THE HOSPITAL LABORATORY: Strategy, Equipment, Management and Economics
T. B. HALES, Arrowe Park Hospital, Merseyside
OFFSHORE PETROLEUM TECHNOLOGY AND DRILLING EQUIPMENT
R. HOSIE, formerly of Robert Gordon's Institute of Technology, Aberdeen
MEASURING COLOUR: Second Edition
R. W. G. HUNT, Visiting Professor, The City University, London
MODERN APPLIED ENERGY CONSERVATION
Editor: K. JACQUES, University of Stirling, Scotland
CHARACTERIZATION OF FOSSIL FUEL LIQUIDS
D. W. JONES, University of Bristol
PAINT AND SURFACE COATINGS: Theory and Practice
Editor: R. LAMBOURNE, Technical Manager, INDCOLLAG (Industrial Colloid Advisory Group), Department of Physical Chemistry, University of Bristol
CROP PROTECTION CHEMICALS
B. G. LEVER, International Research and Development Planning Manager, ICI Agrochemicals
HANDBOOK OF MATERIALS HANDLING
Translated by R. G. T. LINDKVIST, MTG, Translation Editor: R. ROBINSON, Editor, *Materials Handling News*, Technical Editor: G. LUNDESJO, Rolatruc Limited
FERTILIZER TECHNOLOGY
G. C. LOWRISON, Consultant, Bradford
NON-WOVEN BONDED FABRICS
Editor: J. LUNENSCHLOSS, Institute of Textile Technology of the Rhenish-Westphalian Technical University, Aachen, and W. ALBRECHT, Wuppertal
REPROCESSING OF TYRES AND RUBBER WASTES: Recycling from the Rubber Products Industry
V. M. MAKAROV, Head of General Chemical Engineering, Labour Protection, and Nature Conservation Department, Yaroslavl Polytechnic Institute, USSR, and V. F. DROZDOVSKI, Head of the Rubber Reclaiming Laboratory, Research Institute of the Tyre Industry, Moscow, USSR
PROFIT BY QUALITY: The Essentials of Industrial Survival
P. W. MOIR, Consultant, West Sussex
EFFICIENT BEYOND IMAGINING: CIM and its Applications for Today's Industry
P. W. MOIR, Consultant, West Sussex
TRANSIENT SIMULATION METHODS FOR GAS NETWORKS
A. J. OSIADACZ, UMIST, Manchester
MECHANICS OF WOOL STRUCTURES
R. POSTLE, University of New South Wales, Sydney, Australia, G. A. CARNABY, Wool Research Organization of New Zealand, Lincoln, New Zealand, and S. de JONG, CSIRO, New South Wales, Australia
MICROCOMPUTERS IN THE PROCESS INDUSTRY
E. R. ROBINSON, Head of Chemical Engineering, North East London Polytechnic
BIOPROTEIN MANUFACTURE: A Critical Assessment
D. H. SHARP, OBE, former General Secretary, Society of Chemical Industry; formerly General Secretary, Institution of Chemical Engineers; and former Technical Director, Confederation of British Industry
QUALITY ASSURANCE: The Route to Efficiency and Competitiveness, Second Edition
L. STEBBING, Quality Management Consultant
QUALITY MANAGEMENT IN THE SERVICE INDUSTRY
L. STEBBING, Quality Management Consultant
INDUSTRIAL CHEMISTRY
E. STOCCHI, Milan, with additions by K. A. LOTT and E. I. SHORT, Brunel
REFRACTORIES TECHNOLOGY
C. STOREY, Consultant, Durham; former General Manager, Refractories, British Steel Corporation
COATINGS AND SURFACE TREATMENT FOR CORROSION AND WEAR RESISTANCE
K. N. STRAFFORD and P. K. DATTA, School of Material Engineering, Newcastle upon Tyne Polytechnic, and C. G. GOOGAN, Global Corrosion Consultants Limited, Telford
TEXTILE OBJECTIVE MEASUREMENT AND AUTOMATION IN GARMENT MANUFACTURE
G. STYLIOS, Department of Industrial Technology, University of Bradford
PREPARATIVE AND PROCESS-SCALE LIQUID CHROMATOGRAPHY
G. SUBRAMANIAN, Department of Chemical Engineering, Loughborough University of Technology
INDUSTRIAL PAINT FINISHING TECHNIQUES AND PROCESSES
G. F. TANK, Educational Services, Graco Robotics Inc., Michigan, USA
MODERN BATTERY TECHNOLOGY
Editor: C. D. S. TUCK, Alcan International Ltd, Oxon
FIRE AND EXPLOSION PROTECTION: A Systems Approach
D. TUHTAR, Institute of Fire and Explosion Protection, Yugoslavia
PERFUMERY TECHNOLOGY 2nd Edition
F. V. WELLS, Consultant Perfumer and former Editor of *Soap, Perfumery and Cosmetics*, and
M. BILLOT, former Chief Perfumer to Houbigant-Cheramy, Paris, Président d'Honneur de la Société Technique des Parfumeurs de la France
THE MANUFACTURE OF SOAPS, OTHER DETERGENTS AND GLYCERINE
E. WOOLLATT, Consultant, formerly Unilever plc

PRACTICAL USES OF DIAMOND

A. BAKOŃ
Research Centre of Geological Techniques, Warsaw, Poland

A. SZYMAŃSKI
Electronic Materials Research and Production Centre — CeMat'70 S.A., Warsaw, Poland

Translation Editor
P. Daniel

ELLIS HORWOOD
NEW YORK LONDON TORONTO SYDNEY TOKYO SINGAPORE

PWN—POLISH SCIENTIFIC PUBLISHERS
WARSAW

English edition first published in 1993
in coedition between
ELLIS HORWOOD LIMITED

Market Cross House, Cooper Street,
Chichester, West Sussex, PO19 1EB, England

A division
of Simon & Schuster International Group
A Paramount Communications Company

and

POLISH SCIENTIFIC PUBLISHERS PWN Ltd.
Warsaw, Poland

Translated from the Polish by Jerzy Tomaszczyk

Printed in Poland by D.N.T.

British Library Cataloguing in Publication Data

Bakoń, A.
Practical uses of diamond — (Ellis Horwood series in applied science and industrial technology)
I. Title II. Szymański, A. III. Series
553.6

ISBN 0-13-739095-5

Library of Congress Cataloging-in-Publication Data
available from the Publishers

Table of Contents

Preface

Diamond has always fascinated. For the man in the street, the mineral personifies a magical and virtually unattainable symbol of wealth, except for the once (or more) in a lifetime purchase of an engagement ring. Few people realize, however, how important a role diamond plays in present day industry and thus, indirectly, in the life of every one of us.

Our purpose in writing this book was to acquaint the reader with the major areas of diamond use in the world's economy. For reasons of space, we present a highly condensed account ranging from the properties of diamond to many of its applications, although only in a very few cases have we been able to discuss practical uses of diamond in detail.

For the benefit of the reader who may wish to consult additional sources, a list of references is appended at the end of the book.

Andrzej Bakoń and Andrzej Szymański

Acknowledgements

The aim of our work was to present to the reader some idea of the astounding variety in the practical uses of natural and synthetic diamonds, their technical specifications and the techniques employed when preparing them for specific applications.

We, the authors, can hardly claim exclusive knowledge of this very broad area. Thus, in order to draw as full a picture of the subject as it deserves—within the limits of available space—we approached the world's major diamond suppliers and diamond tool manufacturers for permission to use their publications, illustrative material and statistical data, etc.

Permission was granted in every single case, and we wish to thank the following companies for information supplied:

— The Amplex Corporation, Bloomfield, CT—USA
— Advanced Resins Limited, Cardiff, South Glamorgan—UK
— Boliden Finemet, Skelleftehamn—Sweden
— Coborn Engineering Co., Romford, Essex—UK
— De Beers Industrial Diamond Division (Pty) Ltd., Ascot, Berkshire—UK
— Diagrit Diamond Tools Div. of Diamant Boart, Staplehurst, Kent—UK
— Diamant Boart SA, Brussels—Belgium
— D. Drukker and Zn NV, Amsterdam—The Netherlands
— Dr. Fritsch KG, Fellbach—Germany
— E.I. du Pont de Nemours and Co., Wilmington, DE—USA
— Ewag AG, Solothurn—Switzerland

— General Electric Co., Worthington, OH—USA
— Henri Polak Diamond Corp., New York, NY—USA
— Macro Division of Kennametal Inc., Fort Coquitlam, BC—Canada
— Norton (Eastman) Christensen, Celle—Germany
— Ponar—Łódź, Łódź—Poland
— Rubin and Zoon PVBA, Antwerp—Belgium
— J. K. Smit and Sons Ltd., Colwyn Bay, North Wales—UK
— Tyrolit Schleifmittelwerke Swarovski KG, Schwaz Tirol—Austria
— Urbanek, Joh., Frankfurt a. Main—Germany
— Wendt GmbH, Meerbusch—Germany
— Winter and Sohn GmbH, Hamburg—Germany
— Workdiamond, Roncarolo di Caorso—Italy

We also wish to acknowledge the kind assistance of:

— Dr J. S. Sexton, Fairey Arlon Ltd. (formerly of De Beers Industrial Diamond Division (Pty) Ltd.), Ascot, UK and Mr P. Browne, De Beers Industrial Diamond Division, Shannon, Ireland
— Prof. N. V. Novikov, Director, Institute of Superhard Materials, Kiev, USSR
— Mr H. Hickl-Szabo, H. J. Steffens and P. Frenz, General Electric Superabrasives Europe, Dreieich—Sprendlingen, Germany—for help received when collecting data on diamond and other superhard materials and tools
— Prof. H. Żebrowski, Wrocław Technical University, Wrocław, Poland

The authors express their sincere thanks to Mr Paul Daniel, from De Beers Industrial Diamond Division (Pty) Ltd., for his verification of the English text and for many valuable comments and materials which have greatly contributed to the final version of the book.

Diamond trade marks

When reading through this book, on a number of occasions the reader will come across letter or letter-number symbols. These will generally be the registered trade marks of the various kinds of diamond product available on the world market, denoting a particular diamond abrasive or polycrystalline tool material with specific useful characteristics. Of course, the useful properties differ from one kind of product to the next.

As a rule, the trade marks were devised, and are owned, by the various diamond manufacturers and/or suppliers. As a matter of principle, when marking his packaging with a particular trade mark, the supplier guarantees a certain product quality that will be maintained when or if another order is placed with that supplier. The different kinds of product, their trade names and the respective suppliers are listed in Chapter 3 and in Index.

In most cases a trade mark is an abbreviation of a perhaps lengthy name which signifies the structure of the diamond crystal or mass or its application. For example, **MDS** stands for Monokrystaliczny Diament Syntetyczny, which is Polish for monocrystalline synthetic diamond, **MDA** (a De Beers product) = Metal bond Diamond Abrasive, and **NRBT** (also De Beers) = Natural Resin Bond-Treated.

Similar registered trade marks are used for the many polycrystalline diamond (PCD) materials available on the world market.

In this book, the symbols of Soviet-made diamond products have been transliterated into the Roman script.

1

The structure and properties of diamond

1.1 SPATIAL CONFIGURATIONS OF CARBON ATOMS

Carbon occupies a very specific place among the elements. It comes sixth in the periodic table of the elements, which places it in the fourth group. This means that a carbon atom has six protons in its nucleus and six electrons in orbit around it. It also means that in the valence (outer) shell of a carbon atom there are four electrons: two on the energy sublevel $2s$, one on the sublevel $2p_x$ and one on $2p_y$. Thus, for the valence shell of a carbon atom to be completely filled, it would have to have another four electrons. In chemical terms, such a configuration and number of electrons in the valence shell allows, and shows a tendency to, the formation of four chemical bonds.

The formation of chemical bonds between atoms is accompanied by a release of energy which brings about greater stabilization of the system. In the case of carbon atoms, the energy sublevels $2s$, $2p_x$ and $2p_y$ hybridize relatively easily with one another and form hybridized orbitals capable of covalent bonding. However, depending on external conditions, carbon atoms are capable of self-bonding in various spatial configurations, where the chemical bonds among the atoms may be described by means of hybridizations sp, sp^2 or sp^3. The differences in the spatial configurations of atoms and in the character of the bonds between them are responsible for the existence of a number of carbon allotropes with widely differing properties (see Table 1.1).

Table 1.1 – Allotropes of carbon

Allotrope	Bonding	Structure	Cell parameters (nm)		Density (g/cm^3)	Mohs hardness
			a_0	c_0		
Diamond	sp^3	Cubic O_h^7–Fd3m	0.357		3.515	10
Lonsdaleite (hexagonal diamond)	sp^3	Hexagonal $D_{6h}^4-P_3^6/mmc$	0.252	0.412	3.51	10
Graphite 2H	sp^2		0.246	0.670	2.27	1
Graphite 3R	sp^2	Rhombohedral $D_{3d}^5-R\bar{3}m$		1.006	—	—
Cubic graphite	sp^2	Cubic	0.554		—	—
Carbon from Ries crater	sp	Hexagonal P6/mmm	—	—	—	—
α-Carbyne	sp $\equiv C-C\equiv C-C\equiv$	Hexagonal	0.892	1.536	2.68	Between graphite and diamond
β-Carbyne	sp $=C=C=C=C=$		0.824	0.768	3.11	

1.2 THE STRUCTURE OF DIAMOND

Diamond is a crystal allotrope of carbon, the atomic configuration of which is described in terms of sp^2x hybridization while the symmetry of its atomic configuration is characterized in terms of the space group $Fd3m$—O_h^7. Figure 1.1 shows the ideal structure of diamond. The positions of the atoms are: $(0,0,0)$, $(\frac{1}{2},\frac{1}{2},0)$, $(0,\frac{1}{2},\frac{1}{2})$, $(\frac{1}{2},0,\frac{1}{2})$, $(\frac{1}{4},\frac{1}{4},\frac{1}{4})$, $(\frac{1}{4},\frac{3}{4},\frac{3}{4})$, $(\frac{3}{4},\frac{1}{4},\frac{3}{4})$, $(\frac{3}{4},\frac{3}{4},\frac{1}{4})$. Each of the carbon atoms forms four equivalent covalent Σ bonds, the angles between them being 109° 28′. The C—C bond length at 298 K is 0.154450 ± 0.00005 nm, while the lattice constants measured at this temperature are in the range $0.356683 \pm$ $\pm 1 \times 10^{-6}$ nm to $0.356725 \pm 3 \times 10^{-6}$ nm [80, 198]. The density of diamond as calculated on the basis of the lattice constant is 3.51525 g/cm^3.

Real diamonds, i.e. as found in nature and as produced by synthesis, contain other elements in addition to carbon, which may considerably influence the properties of the diamond and, as a result, the uses of the diamond material. The type and concentration of impurities depends on the growth environment and conditions. Of the various elemental inclusions detected in diamond, some twenty five are present in concentrations exceeding 10^{-4} atom. % [148, 161]. In addition to C^{12} the C^{13} isotope also occurs in diamond. According to Orlov [148], the C^{12}/C^{13} ratio in natural diamond varies between 89.24 to 91.56. 'Foreign' atoms may be incorporated in the diamond lattice as single substitutional atoms, or they may occupy interstitial sites. They may be present in the form of small aggregates in specific spatial configurations, or they may occur in the form of inclusions, e.g. of other minerals or gases.

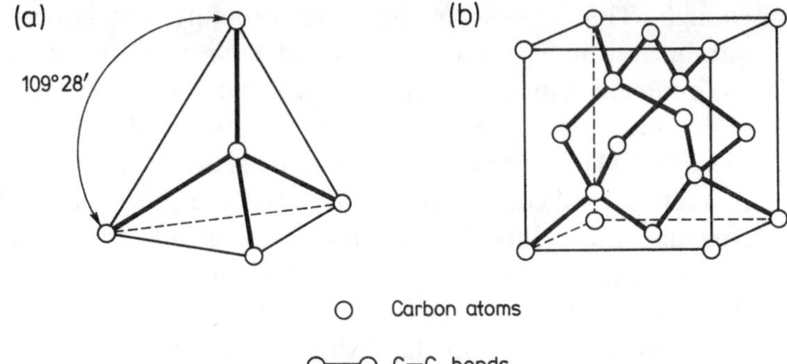

Fig. 1.1 — Ideal structure of diamond: (a) spatial distribution of chemical bonds between carbon atoms in diamond; (b) distribution of carbon atoms and chemical bonds in elementary cell of diamond crystal.

The orientation of impurity atoms and/or inclusions may be compatible or incompatible with the crystal lattice of the carbon atoms.

1.3 ENERGY STABILITY OF THE DIAMOND STRUCTURE

Figure 1.2 is a phase diagram of carbon showing the pressure–temperature conditions of diamond stability. Under normal conditions diamond is not a stable form of carbon in thermodynamical terms, and its crystal lattice may be transformed. The activation energy of diamond transformation to other forms of carbon depends on the internal and external structure of the crystal [11, 80].

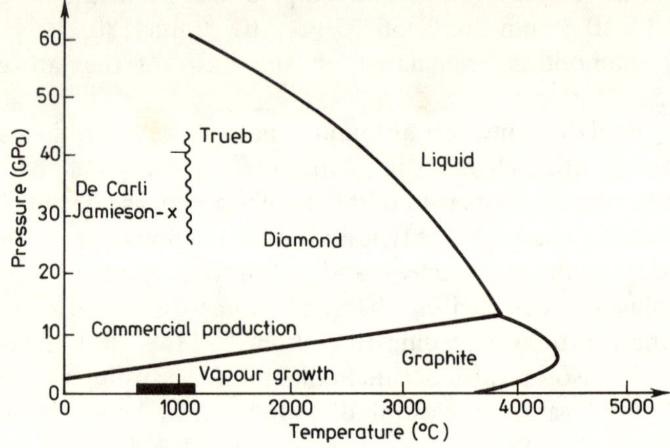

Fig. 1.2 — Phase diagram of carbon with the major pressure–temperature regions in which diamond has been obtained [97].

The diamond structure will transform into graphite even under chemically inert conditions, provided that the temperature is high enough. The process of diamond structure breakdown, known as graphitization, begins at the diamond–environment interface, where by environment is meant the environment outside the crystal, or inclusions within the crystal. Breakdown proceeds into the diamond crystal lattice (see Fig. 1.11). The first to graphitize are sharp crystal edges or the edges of fractures and corners, following which the process extends to surface irregularities: eventually graphitization even affects smooth surfaces. In the course of diamond-to-graphite transformation, the carbon atoms are displaced and the interatomic bonds change their character and orientation. Changes in the distance between crystallographic planes and the thermal-mechanical stresses lead to the appearance of secondary fractures. The carbon product arising from diamond transformation is first and foremost a mixture of various forms of graphite, the $\{0001\}$

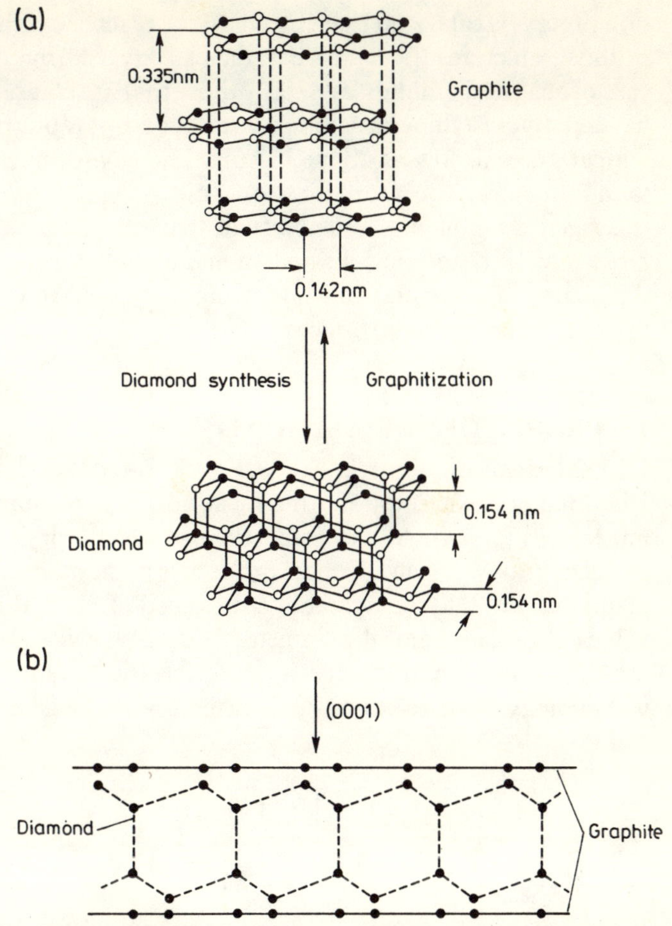

Fig. 1.3 — Comparison of spatial arrangement of carbon atoms in graphite and diamond (a). The graphite axis {0001} is oriented parallel to the {111} axis of diamond (b).

graphite axis going parallel to the {111} diamond axis (Fig. 1.3). Pure natural diamonds and good quality synthetic diamonds begin to graphitize at about 1770 K. Thermally the most stable are pure crystals where freedom from impurities is indicated by the absence of appropriate reflexes on X-ray photographs [20]. Impurity atoms, especially aggregates of such atoms and large inclusions, adversely affect the stability of the diamond structure regardless of whether they occurred naturally or whether they were incorporated during synthesis. The temperature at which graphitization commences decreases in proportion to the number of inclusions. Especially susceptible to graphitization are crystals containing inclusions of metallic elements. The graphitization process is catalyzed by elements which react chemically with carbon or those which readily dissolve it, such as oxygen, iron or nickel. In the case of opaque diamonds, especially those that are black, are fractured or have a polycrystalline structure, degradation by graphitization begins at temperatures as low as about 1020 K. The environment has also been found to affect the graphitization rate [83, 127]. When heated in hydrogen, diamonds — and above all synthetic diamonds — graphitize more slowly than when heated in inert gases (argon, nitrogen), the differences in graphitization rate being attributed to chemisorption of hydrogen on diamond surfaces.

1.4 PROPERTIES OF DIAMONDS

1.4.1 Mechanical properties

Diamond is the hardest and the most abrasion resistant of all known minerals (Fig. 1.4). At the same time it has the highest modulus of elasticity [100]. The mechanical properties of monocrystalline synthetic diamonds are fully comparable with those of natural diamonds. The values of the mechanical parameters vary over wide ranges depending above all on the internal structure of the crystals and on their morphology. As a rule, non-carbon inclusions weaken the crystals.

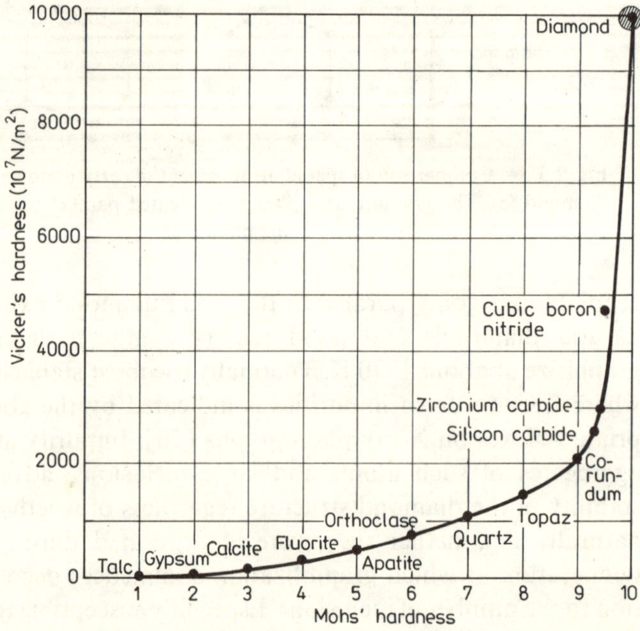

Fig. 1.4 — Diamond hardness related to the hardnesses of other substances.

The carbon atoms in the diamond lattice show different arrangements in different planes, and these are responsible for the anisotropy in mechanical properties of diamond. The arrangement of carbon atoms in the three most characteristic lattice planes of diamond is depicted in Fig. 1.5a. The highest density of atoms is in the octahedral plane {111}. Taking the density in the hexagonal plane {100} as 1, the densities in the {111} and {100} planes are, respectively, 2.308 and 1.414. The directional differentiation of atomic density is responsible for the varying diamond

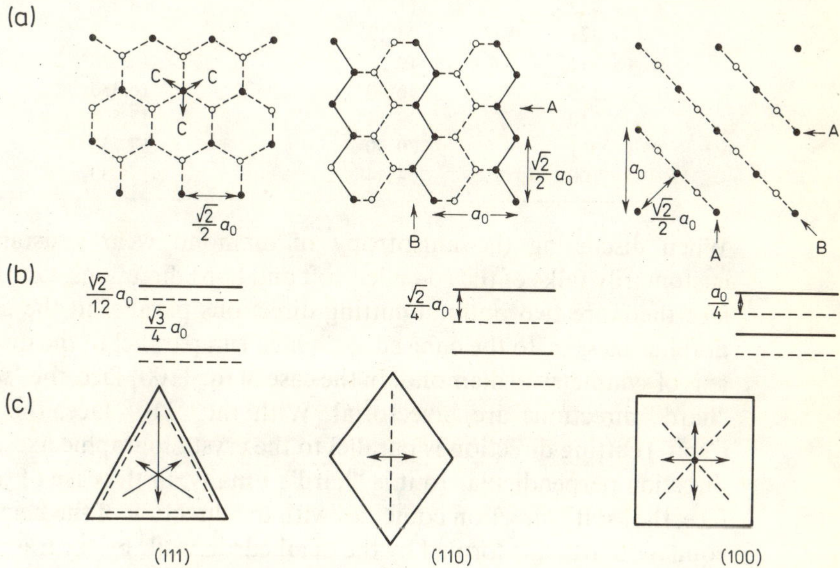

Fig. 1.5 — Spatial distribution of carbon atoms in diamond and the 'hard' and 'soft' hardness directions: (a) distribution of carbon atoms in the {111}, {110} and {100} planes; (b) view of the planes in which carbon atoms are distributed, in the case of sections perpendicular to the relevant lattice planes; (c) directions of highest and lowest abrasion resistance in the three diamond planes. Continuous lines indicate the easiest diamond cutting directions, dashed lines — the hardest cutting directions.

cleavage behaviour in different directions (Table 1.2). Diamond cleaves preferentially in the {111} plane. This fact, as well as the relatively low bending strength of diamond in the {111} direction, must be allowed for in the manufacture and use of many types of diamond tools. Ignoring the cleavage planes may result in premature fracture or tool wear.

Anisotropy of crystal hardness and abrasion resistance is found to occur not only in the different planes but also within the faces themselves, depending on the force direction. Figures 1.5 and 1.6 show the directions of greatest and least abrasion resistance of diamond.

Table 1.2 – Cleavage energy along different crystal planes in diamond [80]

Plane	Angle between plane and {111}	Cleavage energy (erg/cm^2)
111	0	11.330
332	10°0'	12.550
221	15°48'	13.080
331	22°0'	13.510
110	35°16'	13.880
322	11°24'	14.290
321	22°12'	15.730
211	19°28'	16.030
320	36°48'	16.330
210	39°14'	17.560
311	29°30'	17.750
100	54°44'	19.630

When discussing the anisotropy of diamond wear resistance, one customarily talks of the so-called soft and hard directions. On the {100} face there are two optimal cutting directions parallel to the crystallographic axes, i.e. to the cube edges, which run parallel to the quaternary axis of symmetry in diamond. In the case of the {100} face, the "soft" and "hard" directions are bivectorial. With the {110} face, the optimal ("soft") cutting direction is parallel to the crystallographic axes, and the direction perpendicular to it is "hard". Finally, in the case of the {111} face, the "soft" direction coincides with the direction of the height of the equilateral triangle formed by the octahedron wall; by contrast with the cutting directions on the {110} and {100} faces, the "soft" direction on the {111} face is vectorial and runs towards the octahedron vertex.

Grodziński [99] has shown that there are considerable differences between theoretical recommendations and the practice of actual diamond cutting. Thus, while the bivectorial cutting directions on the {100}

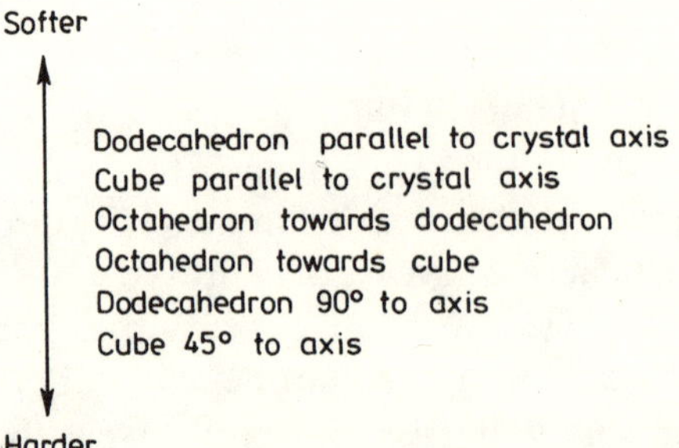

Fig. 1.6 — Dependence of diamond crystal hardness on the crystallographic directions [27, 216].

and {110} faces are theoretically equally abradable in both opposite directions; in practice, one of the directions is distinctly harder. According to E. M. Wilks and J. Wilks [216], this is due to the fact that the crystal faces are out of alignment with the atomic lattice (Fig. 1.7).

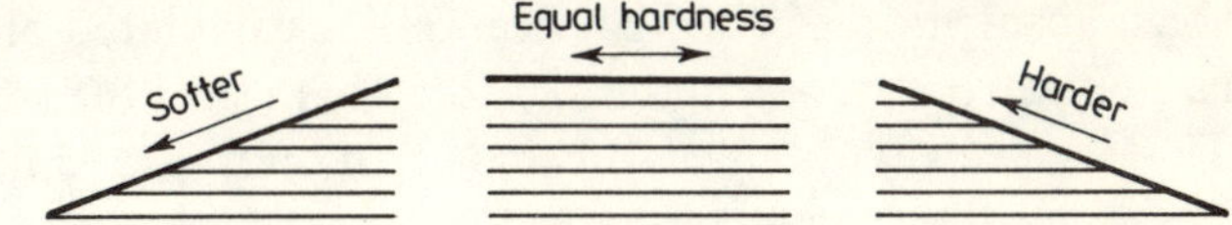

Fig. 1.7 — Polishing "against the lattice" is harder than polishing with it [27].

Misorientation of the face relative to the crystal lattice, perceptible in diamond cutting, may range from a few minutes of arc to as much as several degrees. In grinding "against the lattice", the hardness is greater. The first two softest directions of grinding or polishing (Fig. 1.8) are often the directions of diamond crystal working. When grinding the {111} face of an octahedral crystal, it is advisable to incline it towards the "soft" direction, i.e. {110}. This is done in the working of industrial diamonds, for example, when forming a cutting corner compatible with the quaternary axis of symmetry. After reconditioning, a corner cutting

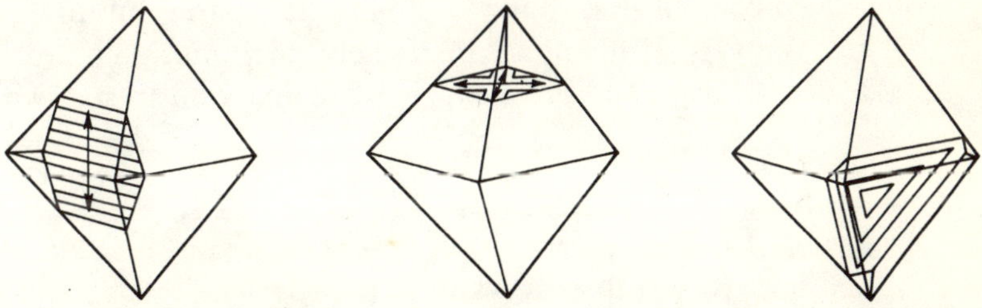

Fig. 1.8 — Possible grinding directions in an octahedral crystal on the dodecahedral and cube faces. The true octahedral face cannot be ground. Here the grinding face is shown tilted, in a direction that permits grinding [27].

point of this kind becomes increasingly softer: since the tool maker cannot regrind the octahedron face, he has no choice but to incline it, if only a little, in a direction more amenable to abrasion. A general recommendation is to orient the "crystal lattice" perpendicular to the direction of grinding wheel travel. When grinding an octahedron in the {100} plane, which is perpendicular to the quaternary axis of symmetry, the lattice lines are parallel to the diagonals (Fig. 1.8). In the {110} plane the lines run parallel to the line connecting two vertices, the lines being univectorial. In a rhombic-dodecahedral crystal, the corners made by three faces are regarded as "soft", and those made by four faces as "hard".

The optimal direction of diamond cutting should run from one "soft" corner to another.

1.4.2 Optical properties

Ideally pure diamond is a colourless mineral. The threshold frequency at which absorption of UV light occurs—as calculated on the basis of the diamond structure—is in the 220–225 nm range [148]. In most cases the UV–IR spectra obtained for red diamonds are quite complex and depend on the specific structural features which, in turn, are related—among others—to where they were mined or how they were synthesized. Thus, side by side with colourless diamonds there are crystals with different shades of yellow, green, brown, violet, blue, pink, as well as milk-white and black, the hue being due to the presence of impurity atoms. Foreign atoms affect the shapes of the spectral lines, the extent of the changes depending on their number and distribution. The absorption of radiation is especially sensitive to the presence of nitrogen atoms: several types of diamond are distinguished in terms of the number of nitrogen atoms and their distribution. These different types respond differently to light and to other external stimuli (Table 1.3).

Typical curves characterizing UV absorption spectra for diamond are shown in Fig. 1.9. According to Orlov, diamonds with very low nitrogen content (less than 0.01%) are transparent up to 225–230 nm. These are type II diamonds, extremely rare among natural diamonds (De Beers, Finsch and Premier Mines). In diamond crystals with a higher

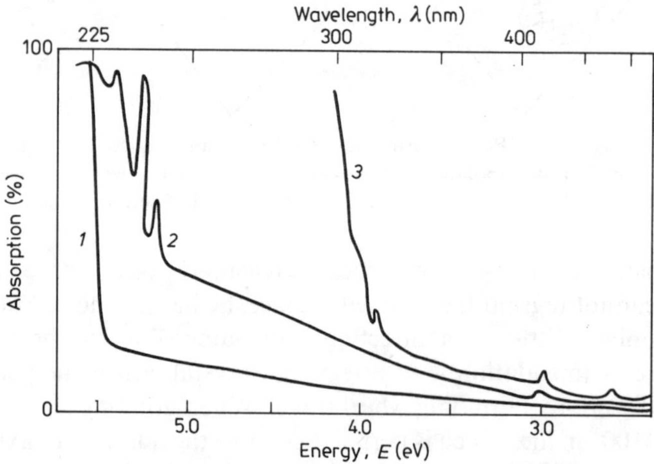

Fig. 1.9 — UV absorption spectra of diamonds: *1* — continuous absorption from $\lambda = 225$ nm in type IIa diamonds (= nitrogen poor), *2* — absorption with fine structure in 225–240 nm regions (N9 system), *3* — continuous absorption from 300 nm in type Ia diamonds [148].

Table 1.3 – Basic physical properties of the two main types of diamond [27, 100]*

	Type I diamond		Type II diamond	
	Ia	Ib	IIa	IIb
Characteristic	Contains nitrogen as an impurity in fairly substantial amounts (of the order of 0.1%), and which appears to have segregated into relatively large sheets or platelets within the crystal. Most natural diamonds are of this type	Also contains nitrogen as an impurity but in dispersed form. Almost all synthetic diamonds are of this type	Effectively free of nitrogen impurity. Very rare in nature. These diamonds have enhanced optical and thermal properties	A very pure type of diamond which has semiconducting properties: generally blue in colour. Extremely rare in nature. Semiconducting properties can be imparted to synthetic crystals by the incorporation of suitable impurities
Physical properties				
Infra-red absorption	Show absorption between 8 and 10 μm		No absorption between 8 and 10 μm	
Ultraviolet	Complete beyond 330 nm		Transparent down to 220 nm	
Photoconductivity	Weak		Strong	
Birefringence	Usually strong		Weak or absent	
Thermal conductivity	Very good		Extremely good	
Electrical conductivity	None		Type IIb are semiconductors	
Luminescence	Present		Present with differences	
Morphology	Well-formed single crystals		Poorly developed	
Cleavage	Relatively uneven		Relatively perfect	

* Type III diamonds — a proposed classification for meteoric diamonds, which were established by K. Lonsdale as having a hexagonal structure, and have been given the name Lonsdaleite.

nitrogen content (up to 0.10%) in the form of associations of atoms called platelets, the continuous absorption edge for UV light lies in the range 300 to 225 nm, the precise position depending on the nitrogen content. An increase in nitrogen content in this form shifts the absorption edge toward longer wavelengths. Diamonds with these optical characteristics constitute an intermediate type. They are considerably more abundant than type II diamonds, although they account for only a small proportion of the total diamond population. Some diamonds previously believed to be of an intermediate type, on the basis of their transparency in UV light show an absorption line system in the 225–240 nm region. It is assumed that these diamonds contain nitrogen associated with dislocation loops in the {111} planes [148]. Diamonds with a relatively high nitrogen content (0.10–0.25%) in the form of N_2 molecules and/or nitrogen enriched microareas — so-called platelets— are opaque to ultraviolet light at wavelengths below 320–300 nm. These are type I ("nitrogen-rich") diamonds. The continuous absorption at wavelengths shorter than 300 nm and the weakly expressed structure at the longwave edge, as observed in type I diamonds, are due to nitrogen impurity forming non-paramagnetic associations of two substitutional atoms. If the diamonds contain nitrogen segregations (platelets), they give rise to progressive absorption in the 250–290 nm region. This type of absorption is observed for low nitrogen impurity content in the N_2 form, which is responsible for continuous absorption from 320 nm. When nitrogen enters the lattice substitutionally, one observes absorption at wavelengths shorter than 500 nm, increasing monotonically at shorter wavelengths [148].

The character of the UV absorption spectrum is closely connected with the spectral characteristics of diamond in visible light and in the IR. All light-blue and blue crystals are transparent up to 225 nm and in structural terms they are designated type II. Yellow and green-yellow crystals do not transmit radiation below 300 nm. Their colour may be due to the presence of isolated paramagnetic nitrogen atoms in place of carbon atoms; it may be related to the presence of N—Al centres, it may result from the presence of double nitrogen atoms associated with vacancies or from the presence—independently—of nitrogen atoms and vacancies. At the same time, absorption of visible light by diamonds is significantly influenced by the presence of other elements. Thus, in most cases, the colour of a diamond depends on the presence of (impurity) ions of iron, titanium, magnesium, aluminium, boron, calcium, etc. Also present in diamonds may be inclusions of olivine, pyrope, and other minerals, especially silicates, while metal carbides are generally present in synthetic diamonds. In colourless diamonds, iron is usually present in insignificant quantities. The presence of trace amounts of iron (especially Fe^{3+} ions) and titanium is responsible for the appearance of peaks at 415, 450 and 480 nm. In diamonds with various shades of blue, one finds boron (about 5 ppm). In diamonds from some mines a relationship has been observed between the colour, crystallographic form, and chemical composition. For example, Sellschop [69] investigated diamonds from the Premier Mine in South Africa and found that their colour, extending from yellow via brown to green, correlates with different amounts of oxygen inclusions, from 34 to 181 ppm, while in the diamonds from the Finsch and Jagersfontein mines, an opposite relationship was observed [175].

The IR absorption spectrum of an ideal diamond, which is a typical homopolar crystal, should show absorption only in the 3–6 μm region due to thermal vibration of the carbon atoms in the lattice (two-phonon lattice absorption) [148]. Absorption in this frequency range is observed in all types of diamond, its value being strongly dependent on the temperature. The spectra for real crystals have more diversified shapes characterized by numerous peaks which indicate specific kinds of bands. The different bands are given their own symbols (Table 1.4). Similarly as

Table 1.4 – Wavelengths of infra-red absorption peaks in diamond [3]

Diamond type	Wavelengths of peaks (cm^{-1})
A	1282*, 1203, 1098, 480
B	1426, 1372, 1332, 1175*, 1003, 780, 328
B_1	1331, 1135*, 1100, 1010
B_2	1430, 1365*, 330
C	
Natural	1129*, 1250–1339, 1345
Synthetic	1100, 1135*, 1345
D	990, 1065, 1290–1320, 1335

* Main peak.

in the case with other frequency ranges of electromagnetic radiation, the character of absorption lines depends strongly on nitrogen inclusions and on the way in which they are incorporated in the diamond crystal lattice. In the case of some frequencies, the absorption coefficients may be represented as a linear function of nitrogen atom content [67]. For example, at a wavelength of 7.8 μm (1280 cm^{-1}) this dependence is expressed as $N = 5.8 \times 10^8 K$, where N is the content of nitrogen atoms, and K is the absorption coefficient. Similar linear relationships exist for other frequencies, including those in the UV region, e.g. for 306.5 nm. Relationships of this kind are observed especially in the case of type I diamonds with a high nitrogen content. With type IIb diamonds they are observed at 2465 and 2810 cm^{-1}.

Diamonds may change their spectral characteristics when heated, compressed, irradiated, bombarded with particles (e.g. neutrons, electrons) etc. or as a result of diffusion or implantation of foreign atoms in the carbon crystal lattice [77]. Subjected to such influences, diamonds may show photoluminescence, X-ray luminescence, cathodoluminescence, thermoluminescence or electroluminescence effects, as well as radioactivity and triboluminescence. The extent and intensity of these effects will depend—as with other optical properties of diamonds—on the internal structure of the crystals [80, 148]. An important role in the formation of luminescence centres is played by paramagnetic nitrogen, while the incorporation of boron permits the formation of the non-paramagnetic association B—N. The changes in diamond properties brought about by external influences may be temporary or permanent. Also, the effects may take place throughout the bulk crystal or only in its outer layer (Table 1.5).

Table 1.5 – Colour effects produced by irradiation [77]

Diamond type / Treatment	Ia	Ib	IIa	IIb
Neutron irradiation	Green	Green	Green	Green
Neutron irradiation + heating	Yellow-amber	—	Brown	Red-purple
Electron irradiation	Green	Blue or greenish-blue	Blue or greenish-blue	—
Electron irradiation + heating	Yellow-amber	Red-purple	Brown	—

Present-day techniques of diamond "tinting" make it possible to obtain bluish crystals, e.g. by bombarding them with a flux of electrons, that are almost indistinguishable from naturally bluish diamonds even though the change of colour is often restricted to the outer layer. The two can, however, be readily distinguished by examining their absorption in the UV and IR. To protect the customers, gemstone retailers in practically all countries are obliged to state exactly what they are selling, i.e. what the true nature of a crystal is and, in particular, whether the stone may change its colour in a reversible way.

Diamond has an exceptionally high refractive index. At the same time diamond is characterized by a very large difference in refractive indices for red and violet light, such high dispersion being quite rare among minerals or man-made crystalline phases (220). Among precious stones only benitoit, ($BaTiSi_3O_9$), shows an equally high dispersion while zircon ($ZrSiO_4$), shows a similar, but not quite as high dispersion. On the other hand, titanite ($CaTiOSi_4$), demantoid, $Ca_3Al_2(SiO_4)_3$, and rutile (TiO_2), as well as silicon carbide (SiC) and yttrium aluminium garnet (YAl_2O_4), show higher dispersion than does diamond. Because of the high refractive index, total reflection of light takes place at an angle of incidence as small as 24.5°. At the same time thanks to the high dispersion of light, one observes in diamond a most beautiful play of colour of the reflected rays.

Crystallizing in a regular system, diamond shows no double refraction [129, 148]. However, double refraction may occur as a result of internal reflection of light rays from mineral inclusions. In such a case, one observes in a polarizing microscope interference effects characteristic of birefringent bodies.

1.4.3 Thermal properties

The thermal conductivity of diamond is compared with the conductivities of some other materials in Fig. 1.10. Diamond is the best heat conductor of any known material at temperatures typical for human life [68]. For example, the thermal conductivity of diamond at room temperature (293 K) is 26W/deg·cm which is several times higher than that of copper [3]. For that reason diamonds are employed in some electronic devices as heat sink elements. The precise value of the thermal conductivity of diamond depends on the crystal structure. The best heat

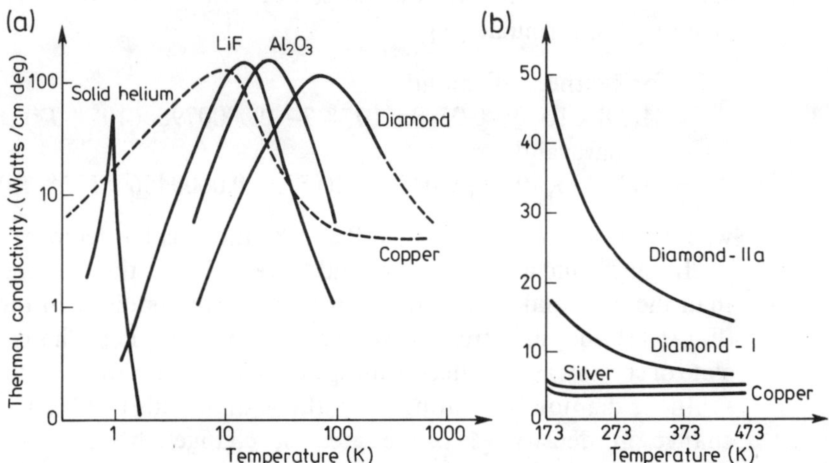

Fig. 1.10 — Thermal conductivity of diamond crystals: (a) thermal conductivities of diamond, copper, solid helium, LiF and Al_2O_3; (b) thermal conductivities of type II and type I diamond and those of silver and copper [21, 68].

conductors are generally crystals of type II diamond, i.e. the most pure diamonds which contain the lowest concentration of nitrogen impurity. The thermal conductivity changes with the temperature, the maximum value being 120W/deg·cm which occurs at about 83 K. Above that temperature, the thermal conductivity of diamond decreases but it remains higher than the values for other known industrial materials.

The thermal conductivity of diamond also decreases when the crystal is irradiated with high energy particles [21]. Rezkov *et al.* [148] have demonstrated the existence of anisotropy in the thermal conductivity of diamond. The isothermal planes in diamond crystals can be represented as triaxial ellipsoids. Thermal conductivity values of diamond calculated in the direction of the ternary axis are 1.9–3.7 times lower than those calculated for the direction of quaternary axis. Because of its excellent thermal conductivity, diamond appears cold when touched by the tongue. This property of diamond makes it easy to distinguish it from 'paste': being a bad heat conductor, the glass 'paste' does not produce the sensation of cold. When breathed upon, diamond mists up more readily than does glass, and the mist disappears from diamond more quickly than from glass. The high thermal conductivity of diamond also lies at the basis of instruments used by jewellers to detect 'simulants'.

At 293 K the specific heat of diamond is 516–550 J((kg · K) [3]. The thermal expansion coefficient calculated by X-ray methods from lattice measurements is from $1.3 \times 10^{-6}\ K^{-1}$ at 298 K to $7 \times 10^{-6}\ K^{-1}$ at 1673 K [3]. The thermal expansion of natural and synthetic diamonds at 293 K is similar at $1.0–1.2 \times 10^{-6}\ K^{-1}$. The values for the thermal expansion coefficient over the temperature range 273 to 773 K may be given by the formulae [3]:

—for synthetic diamond,

$$= A(2.8 \times 10^{-6} + 0.0598 \times 10^{-6}\ T - 0.0000795 \times 10^{-6}\ T^2)\ (K^{-1})$$

—for natural diamond,

$$= A(3.333 \times 10^{-6} + 0.0426 \times 10^{-6}\ T - 0.0000456 \times 10^{-6}\ T^2)\ (K^{-1})$$

where A is a constant equal to 0.28038, and T is the temperature.

Impurity inclusions in diamond have different thermal coefficients than the diamond lattice, and heating will cause stresses in the crystal (Fig. 1.11) and, in extreme cases, crystal fracture [11]. Changes in the state of crystals as a result of heating are observed especially in the case of synthetic diamonds. Heating of synthetic diamonds to 1473 K causes a change in density [41], the greatest changes being observed for polycrystalline diamonds. During heating, small metallic inclusions combine to form larger clusters, most frequently in the {111} plane, resulting in extensive discontinuities in the internal crystal structure.

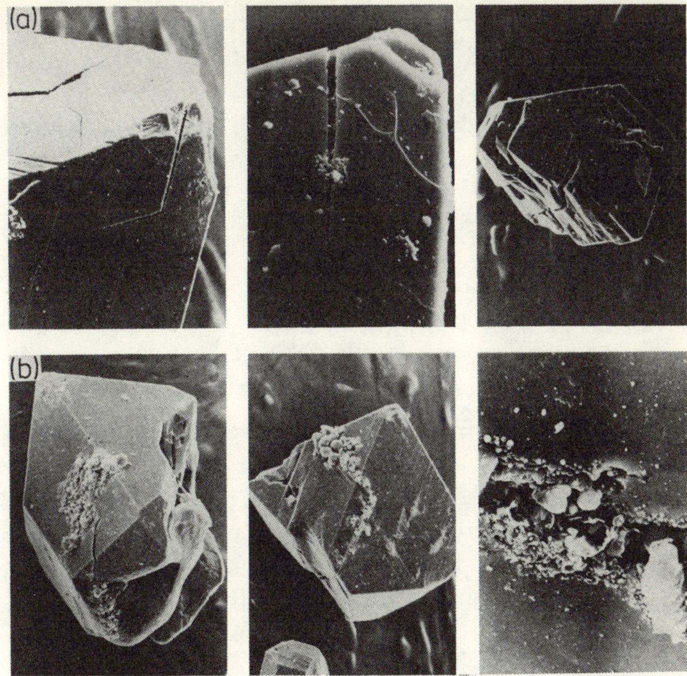

Fig. 1.11 — Destructive effect of elevated temperatures on synthetic diamond grain: (a) grain fracture; (b) graphitization at fracture edges [10].

Generally speaking the more impure the crystal, the greater the effect of heating.

1.4.4 Chemical properties

It is highly characteristic of diamond that it does not react to common acids even at elevated temperatures [80], a feature which differentiates diamond from graphite. The different behaviour of the two crystalline forms of carbon makes it possible to separate them chemically, e.g. from synthesis products. Treated by a hot chromic acid cleansing mixture or a mixture of sulphuric and nitric acids, graphite slowly oxidizes while diamond is chemically inert. On the other hand, diamond oxidizes (graphitizes) relatively readily at high temperature in an oxygen atmosphere and in air. Also, molten hydroxides, the salts of oxy-acids and some metals have a corrosive effect on diamond.

The chemical reactivity of diamond depends on the state of its facets. The temperature of the onset of diamond oxidization is significantly affected by the magnitude, habit and development of the crystal facets. Adsorption of oxygen on diamond facets commences at a temperature as low as 195 K. The heat of oxygen chemisorption on the facets of natural micron diamonds having total facets of 20 m^2/g weight of diamond increases linearly from 6.2 to 59.0 kJ/mole with a simultaneous temperature rise from 301 to 414 K and a pressure increase from 14 to 133 Pa [80]. Oxygen chemisorption proceeds most rapidly on the {111} planes. Above 473 K, a layer of oxide forms on the diamond facets from which — starting at about 623 K — carbon oxides evolve, a process which becomes more intensive from about 653 K. Complete desorption of the oxide layer occurs at about 800 K. According to Sappok and Boehm [174], adsorption of oxygen and water vapour on diamond facets leads to the formation of active groups: carboxyl (—COOH) on the diamond crystal points and edges, carbonyl (=C=O) on the {100} planes, as well as ≡C—OH and ≡C—O—C≡ groups on the {111} and {110} planes; complete coverage of the crystal facets with adsorbed oxygen is possible at 693 K. Natural diamonds of blocky structure and with rounded edges and corners are relatively more resistant to oxidation than polycrystalline stones with more developed facets. Crystals containing a high concentration of metallic inclusions (i.e. synthetic diamonds) break down at lower temperatures, even though the presence of some substances increases the resistance of diamonds to oxidation [182]. For example, in the course of diamond synthesis in the Ni—Mn—C system, the presence of γ-Fe_2O_3 improves the resistance of diamond monocrystals to oxidation. In a similar way, the introduction of 5% boron atoms into the crystal lattice raises the temperature at which diamond oxidation commences by about 75 K.

Evans and Phaal [160] have demonstrated that the facets in the {111} and {110} planes of diamond become coated with graphite when heated from about 973 K, while the {100} face becomes affected only at about 1123 K. Thus, the rate of oxidation with gaseous oxygen, which proceeds via the graphitization stage, is lower on the {100} face than on the {111} and {110} faces.

At temperatures above 870 K, diamond reacts with water vapour and with CO_2 [80]. The rate of natural micron diamond oxidization in CO_2 is higher than for synthetic micron diamond. Strong oxidizers, such as chlorates and perchlorates, cause characteristic triangular etch figures called trigons to be formed at temperatures as low as about 650 K [80], whilst the same applies to hydroxides and alkali metal nitrates. The most aggressive of the nitrates is potassium nitrate or, strictly speaking, the products of its decomposition above 870 K. The rate of diamond oxidation by liquid salts containing potassium is twice as high as the rate of etching by sodium-containing salts [171]. The rate of etching of diamond by liquid hydroxides and salts is highest in the {111} plane and lowest in the {110} plane, with an intermediate value for the {100} plane [136].

With metals, diamond may chemically react with them and form carbides, or it may dissolve in the metal. The metals which at high temperatures form carbides include tungsten, titanium, tantalum and zirconium, while those which dissolve diamond are iron, cobalt, nickel, manganese and chromium. In such molten metals, carbon usually dissolves up to several per cent, but sometimes it may also form carbides [80]. The fact that diamond dissolves in and/or reacts with iron or iron alloys (e.g. steels) above 950 K makes diamond tools unusable for most machining operations on ferrous metals, including high speed and hardened steels [80]. In the grinding and machining of such materials, diamond (and carborundum) has been replaced by cubic boron nitride in abrasive or polycrystalline mass form.

1.4.5 Electrical properties

Like the other elements of the IV group in the periodic chart which form monoelemental crystals with a diamond structure, such as silicon or germanium, carbon with this structure has dielectric properties. The width of the forbidden zone ΔE is equal to 5.7 eV and the calculated resistivity for perfect diamond is 10^{70} Ohm·cm [148]. In reality, the electrical resistivity of crystals depends on the volume of inclusions, their nature and on the way in which they are incorporated into the crystal lattice. In most natural diamond crystals, classed as type I, the resistivity attains values ranging from 10^{14} to 10^{16} Ohm·cm, while type II diamonds have resistivities ranging from 1 to 10^{8} Ohm·cm. Low nitrogen natural diamonds, designated type IIb, have been shown by Custers [148] to be *p*-type semiconductors. A characteristic of natural diamonds with semiconducting properties is their blue colour. Modern diamond synthesis technologies as well as the technologies for doping the internal structure of existing diamonds make available a wide range of diamond semiconductors. The required character and magnitude of conduction can be obtained by electron bombardment or by irradiation with nuclear particles or ions. An acceptor type of conductivity can be imparted to diamonds by using boron and aluminium atoms. *n*-Type conductivity is obtained by introducing such donors as nitrogen, phosphorus, arsenic, or antimony atoms [165].

The conductivity of synthetic diamonds, especially if doped, shows anisotropy [3, 80]. For example, the conductivity of diamond crystals doped with boron is 10^{6}–10^{8} times higher in the {111} direction than in the {100} direction. This is attributed to the different spatial distribution of boron atoms in the crystal structure.

The dielectric permittivity ε of diamond at 300 K is 5.7 ± 0.05 and its temperature dependence can be described [84] by the equation:

$$\varepsilon = 5.70111 - 5.35167 \times 10^{-5}\ T + 1.6603 \times 10^{7}\ T^2$$

Irradiated diamonds exhibit photoconductivity. The photoemission threshold (electronic work function) amounts to 6 eV, which corresponds to a UV wavelength of 207 nm [80]. However, a photocurrent can be recorded using UV irradiation of 210–310 nm wavelength, the value of the current obtained with type IIa diamond being higher than under the same conditions with type I diamonds [80, 148]. The effect of fast particles, e.g. of ionizing radiation, on photoconducting diamonds, especially type II, is such that they result in the formation of current pulses. This property makes it possible to employ specially selected diamonds in conduction counters. Diamonds have also been used in scintillation counters.

2

Natural diamond sources and methods of synthesizing diamond

There are two ways in which diamond can be obtained. The traditional way is mining, but occurrences are extremely rare. The other way, available since 1953, is the manufacture of carbon materials with the structure of diamond using methods developed by man. These different modes of diamond recovery underlie the now traditional division into natural and synthetic diamonds.

2.1 NATURAL DIAMONDS

2.1.1 Morphology and structure of natural diamonds

Natural diamond crystals exhibit considerable variation in size and structure. Large diamonds are found only very infrequently. According to Williams [217], in 50 years of mining South African deposits (1880–1930) three crystals were found of over 1000 carats (including the Cullinan), 11 crystals of 500–1000 ct, and 702 of 100–500 ct.

Some properties or features of diamonds are traditionally associated with the source, for they are present only in or on diamonds found in a particular area. For example, the colour designation 'jager' comes from the name of a South African mine, and the structure designations 'Congo' or 'West African' are obviously related to mining areas.

Fully and perfectly formed crystals which could be used in tools or as jewels without cutting, cleaving or some other kind of treatment are extremely rare. Even crystals characterized by similar habit show differences in facet sculpture or internal structure. The most common flat-faced form of diamond is the octahedron. Well formed octahedra with flat, smooth faces and straight edges are commercially known as 'glassies'. Crystals of this kind are however very often deformed as a result of non-uniform growth. As a rule, they exhibit good transparency or clarity. Non-uniform crystal growth in the different directions sometimes leads to tabular habit. In such a case thin crystals, known as flats, which have parallel octahedral faces, well developed in one direction and poorly developed in the others, result (Fig. 2.1).

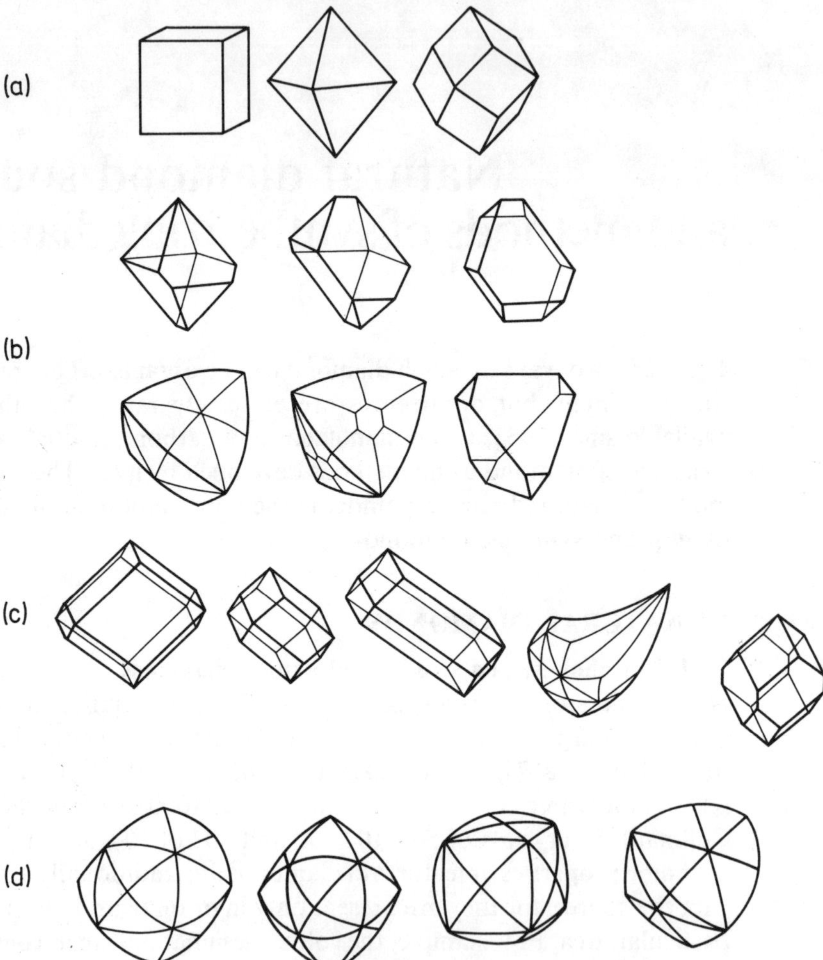

Fig. 2.1 — Examples of natural diamond crystal development: (a) characteristic flat-faced species: cube, octahedron, rhombic dodecahedron; (b) distorted octahedra (tabular at top right); (c) other forms of distorted crystals (dodecahedra and hexakis octahedron); (d) examples of spherical diamond crystals: dodecatroid, octahedroid, cuboid and tetrahedroid [27].

More common are diamond crystals which combine octahedral forms with some other forms. The faces of such crystals are often convex and rough. The roughness—in the form of furrows, accretions, hummocks and pits—results from irregular crystal growth or from post-growth contact with the environment. On the faces one can distinguish steps, terraces and features with triangular contours, known as trigons, which may be concave or convex.

Cubic crystals are quite common but, because of their poor quality, they are of little use as gems. Typical cubic forms are rounded, often domed, with accretions or etch features which form terraced steps along the edges or pyramidal pits along the face diagonals [27]. The irregularities appearing on $\{100\}$ faces and related to imperfect growth have the form of terraces with square contours.

Fully and perfectly developed flat-faced rhombic dodecahedra are rare. More common are ovalized forms (dodecahedroid) or intermediate, modified forms, combinations of octahedra and dodecahedra, which sometimes have the shape of a flattened pillow, or are elongated or plate-like. In extreme cases the excessive growth of some of the dodecahedron faces may produce long, thin and slightly distorted crystals, sharp at one end and blunt at the other. These are known as 'canine teeth' or 'nails'.

The most common forms of natural diamond crystal are however intergrowths and forms which require specialist gemmological apparatus for their correct identification. A significant proportion of crystals from any mine will have one or more faces or edges damaged. Intergrowths may be made of up of two or more diamond crystals. A characteristic feature of intergrowths is the variation in their physical properties, e.g. hardness, abrasion resistance, on going from one component crystal to another within the same stone. Quite common are twin crystals, especially interpenetrant and contact twins, much less frequently star twins. Interpenetrant twins occur when two crystals grow in the same place but have different crystal orientations. As one crystal penetrates into the other, they have a volume common to both. Contact twins resemble crystals which adhere to each other side-to-side, but they have different orientations. A specific form of diamond contact twin is the so-called macle (Fig. 2.2). Such twins are triangular, usually very thin, often with rounded faces, and sometimes resemble triple or six-fold octahedroids. Among diamond crystals one also finds layered twins, as well as aggregates made up of a number of crystals with different degrees of adhesion between the various crystals. Granular and opaque concentrations, grey or black, are called 'bort' [148]. However, there is a tendency in the gem diamond trade to describe all non-gem diamond as 'bort'.

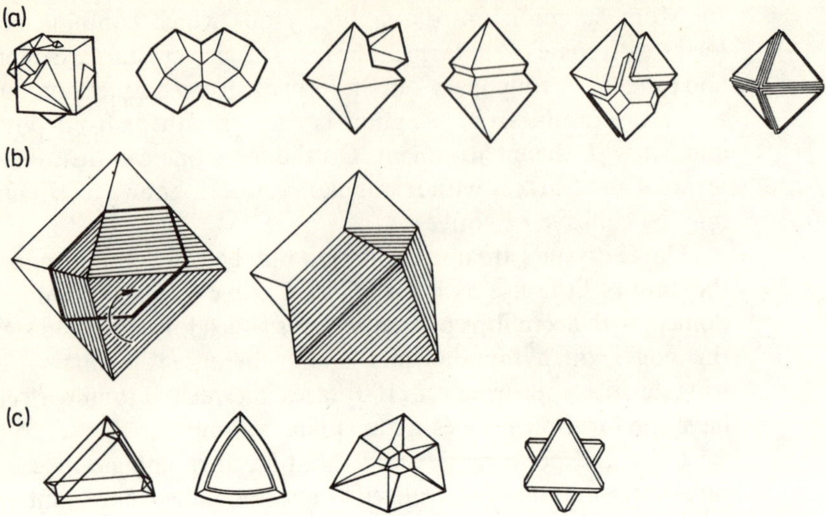

Fig. 2.2 — Examples of simple natural diamond crystal concretions and twins: (a) forms of twinning in diamond: interpenetrant cubes, contact twin dodecahedra, interpenetrant octahedra-parallel growth in octahedra and lamellar twinning in an octahedron; (b) macle formation from rotation of one half of an octahedron; (c) forms of macles left to right: with re-entrant angles, triangle, with hexakis hexahedral faces, and partly rotated to from a star [27].

Apart from 'macro' defects related to poor crystallization, one finds in diamond crystals various kinds of mineral inclusion. The most important of these are [51, 80, 148]:

— olivine, in the form of colourless or pale-green semi-spheres resembling bubbles of liquid. Olivine crystals may occur singly or in clusters, and they are often aligned in octahedral planes parallel to the edges;
— garnet, in the form of pyrope-composition crystals, of brown, orange, yellow, pink, violet and purple hues. Garnet is fairly common in South African diamonds;
— graphite, appearing as black inclusions;
— pyrrohotite, pyrite, pentlandite, ilmenite or rutile, which resemble graphite inclusions (especially common in Ghanaian diamonds);
— diamond inclusions, in a host diamond crystal;
— chrome diopside, appearing as emerald green, well-developed crystals. Like chrome enstatite, these inclusions are fairly common in diamonds mined in the Soviet Union and South Africa;
— chrome spinel in the form of deformed octahedrons, dark brown or black in colour—common in Soviet diamonds;
— 'cloud-like' inclusions, which resemble the Maltese cross (found in Indian diamonds).

Apart from diamond mono- and polycrystals, intergrowths and twins, where the single diamonds may have appreciable size and a distinct structure there occur in nature diamonds with a cryptocrystalline structure Jeynes [111] has identified several types of such diamonds (Tablė 2.1). The most common of these are 'carbonado' and 'ballas'.

Table 2.1 – Different types of natural diamond exhibiting a polycrystalline structure [111]*

Name	Definition	Synonyms
Carbonado	A porous, randomly polycrystalline diamond aggregate. General term for all black irregular diamond. Irregular interlocking crystallites, less than 20 μm	Carbon, Black diamond, Common bort
Stewartite	Magnetic carbonado	Magnetic bort
Framesite	Phanero-crystalline carbonado (grains visible to the eye)	
Ballas	Polycrystalline diamond of oriented globular growth. The crystallites have 110 directions radial	Round bort, Spherical bort, Shot bort
Hailstone bort	Concentric shells of clouded crystal diamond alternating with grey or grey black porous paste	White bort

* Other types are: Oriented carbonado, Coated crystals, Fibrous cubes.

Carbonado is a porous micropolycrystalline mass made up of diamond crystals, the dimensions of which range from a fraction of a micrometer to about 20 μm, but mostly from 1–4 μm. Carbonado occurs as irregularly shaped and fairly large 'chunks'. The largest carbonado piece, found in Brazil, weighed 3167 ct. Ballas-type diamonds are structurally similar but they are less common. They occur in the form of spherical aggregates, are harder and usually mechanically stronger (tougher) than carbonado.

2.1.2 Types of natural diamond deposits

Diamond is found in nature in primary and secondary deposits. These deposits are found only in areas of ancient geological platforms and shields (Fig. 2.3). The chemical and mineralogical composition of diamond ferous rocks all over the world as well as their structure have much in common.

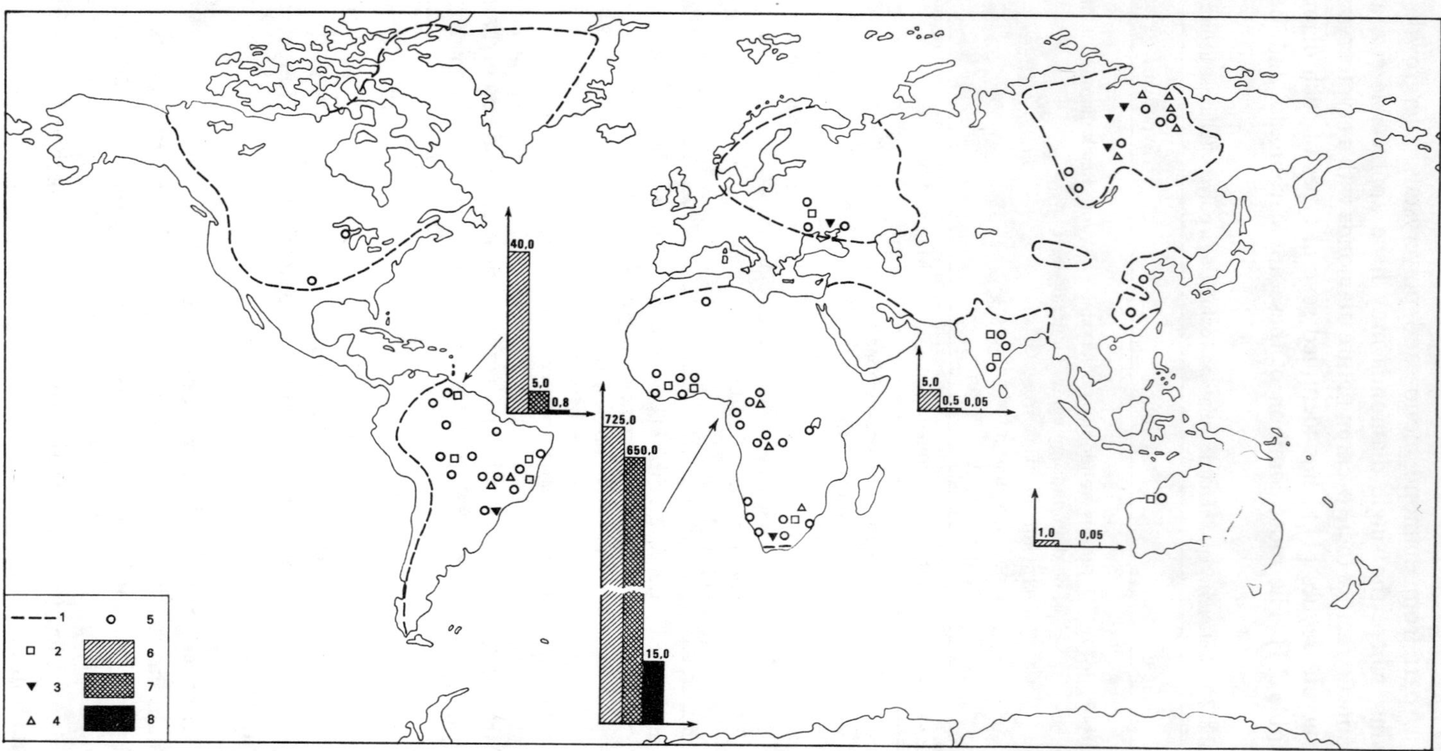

Fig. 2.3 — World distribution of diamond-bearing rock formations and of the secondary deposits related to them: *1* — boundaries of old geological plates, *2–5* — deposits related to Precambrian (*2*), Paleozoic (*3*), Mezozoic (*4*), and Genozoic (*5*) diamond-bearing formations, *6* — estimated diamond reserves (million ct), *7* — estimated diamond production since earliest times (million ct), *8* — estimated annual diamond mining (million ct) [163].

Note: Australian production now running at ~35 million ct p.a. (Authors).

Economically important primary sources of diamond occur in kimberlite rocks, usually connected with peridotite intrusion. Diamonds are also found in peratotites, eclogites and in serpentine pyrope peridotites in the form of xenoliths [27, 51, 53]. Lamproites are another source. The transformation of (unknown) carbon materials into diamonds generally took place under ultra high pressure and high temperature conditions in molten ultrabasic magma in the peridotite mantle of the Earth (Fig. 2.4). It is assumed that kimberlite represents the magma in which diamonds were transported. This magma had a higher iron and magnesium content than the average levels of these elements in the Earth's crust, and it contained very small amounts of quartz.

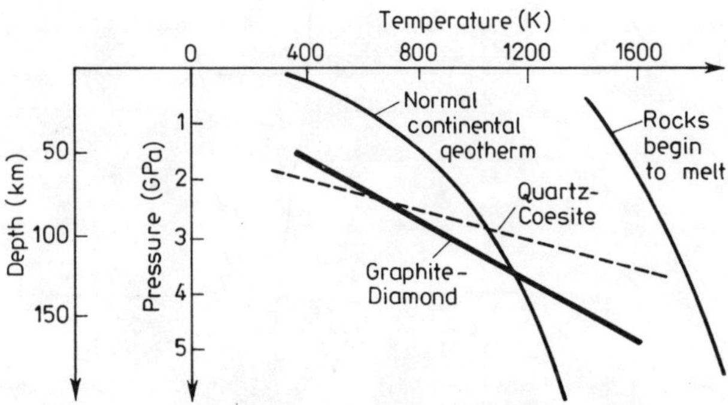

Fig. 2.4 — Diamond stability in the earth [111].

Prospecting for diamond is far more difficult than looking for a needle in a haystack. According to Prokopchuk [163], only about half of the known kimberlite pipes in the USSR contain diamonds, and less than 10% of these contain enough diamond for mining to be an economic proposition — assuming of course accessibility by road or rail. Kimberlite pipes vary greatly in their dimensions, shapes and structure (Fig. 2.5). Thus, the Williamson pipe at Mwadui in Tanzania, the largest discovered thus far, occupies an area of 140 hectares. Nevertheless, the average size of kimberlite pipes ranges from as little as 0.1 to a few hectares. Pipes are frequently found in groups. In most cases the cross-section has a shape approaching a circle or an ellipse, the cross-section area usually becoming smaller with depth. Sometimes parts below the surface are in the form of numerous fissures. As a rule, the quantity of diamond found decreases with increasing depth.

The colour of the kimberlite rock varies with depth. The weathered rock near the surface is yellow (so-called 'yellow ground') and shows poor cohesion, which makes it easy to work. At greater depths, kimberlite changes its colour to greyish-blue ('blue ground') and

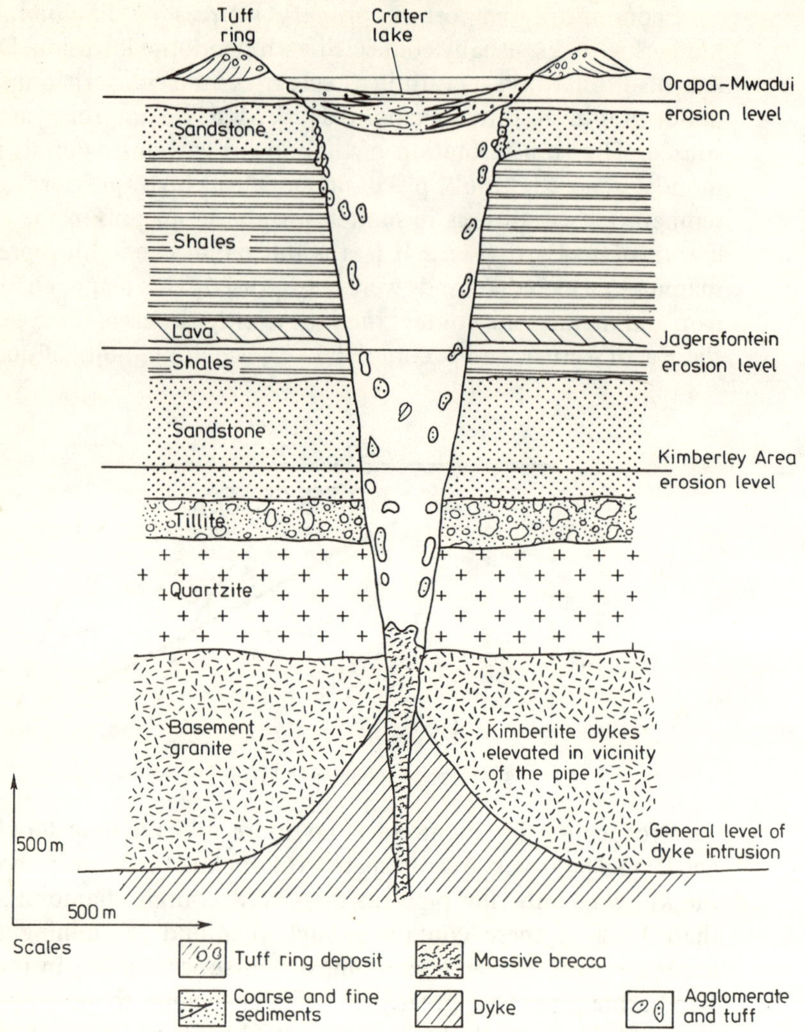

Fig. 2.5 — Generalized model of a kimberlite diatreme and its sub-diatreme dykes [53].

becomes harder and more cohesive. The content of diamond in mined pipes (the 'grade') ranges from a fraction of a carat to several carats per tonne of the rock. Quality of the diamond recovered is, however, far more important than the 'grade'.

Secondary diamond deposits have come about as a result of weathering and erosion of diamond ferous rocks, most frequently thanks to the action of river, sea or rain water. It is estimated [163] that some 70–80% of mined diamond by volume comes from secondary deposits. Economically the most important of these are alluvial and deluvial deposits, with an average diamond content ranging from 0.2 to 2 ct/m^3 of treated gravels and sands. Alluvial deposits were formed by the washing-out of kimberlite pipes followed by transport of diamonds elsewhere by water flow. The width, length and depth of an alluvial deposit depends to a large extent on the water flow speed and on the width of the river or stream, as well as the shape of the river bed and its tributaries. Some well-known alluvial diamond deposits include those in the valleys of the rivers Vaal and Orange in South Africa, and the Kasai, Lubango and Bushimaja in Zaire.

The mining of deluvial deposits, i.e. those formed by the washing-out of primary deposits by rain water, has been reported as accounting for some 50% of world industrial diamond production [163]. Economically important sources of this kind usually contain 0.1–0.3 m ct of diamond and only infrequently about 1 m ct. An exception is the Bakwanga deposit in Zaire, whose reserves have been estimated at 400 m ct.

A highly specific kind of secondary diamond deposit is the undersea deposit. Such deposits are mined on the continental shelf off the Atlantic coast of Namibia and along the coastline. The ratio of diamond to sand or gravel is extremely low, but the quality of the gem diamonds is high enough to make mining economically attractive.

Polycrystalline carbonado and ballas diamonds were formed under geologically quite different conditions. Such diamonds are not found in kimberlite or in secondary deposits formed by the weathering of kimberlite. Since the composition of inclusions in carbonado and ballas diamonds is different from that of the constituents of kimberlite, the origin of these two forms of diamond is thought to be related to metamorphic processes in the Precambrian era [148, 163].

Trace amounts of micron size diamonds are found in some ferro-nickel meteorites [148] and in the sands of some rivers. Several scientists believe that micron size diamonds exist in the cosmos, or extra-terrestrial space.

The mode of diamond mining depends on the pipe structure. Thus, kimberlite pipes are initially mined by the open cast method. This is the method employed at the Finsch and Koffiefontein pipes in South Africa, the Orapa pipe in Botswana, and the Williamson pipe in Tanzania. The deepest excavation of this kind is the 'Big Hole' at Kimberley in South Africa which extends 240 m into the earth (it is no longer worked). Underground deep level or stope mining becomes economical at depth, but underground operations are employed only in some of the South African pipes, predominantly in the Kimberley area, to date. The maximum depth of diamond mining may be 900 m. Mining for gold goes much deeper.

The recovery of diamonds from natural deposits is a major operation involving the transport and processing of huge masses of sand, gravels or kimberlite rock. Linari-Linholm [122] estimated that in the 2000 years of diamond mining, some 230 tons of the precious mineral have been obtained. This necessitated the transport and treatment of 5 million tonnes of ore.

The first step in treatment consists in crushing the ore, which is then sent through hydrocyclones, heavy media separators and flotation chambers. The diamond and water concentrate thus obtained next goes to vibrating grease tables and belts, where diamonds adhere to the grease. Separation may also be carried out by electrostatic, magnetic, and X-ray methods sensitive to differences in the radioactive properties of the materials in the diamond concentrate. In the final stages of treatment use is also made of hand sorting for diamond size and quality.

2.1.3 Commercially worked diamond mines

The major producers of natural diamond today are listed in Table 2.2. India and other South Asian regions, which used to be the main suppliers of gem diamonds in past centuries, now occupy a marginal position. The main diamond ferous areas in India are in Majhgawan and Panna, and in the Mahandi and Krishna river valleys. There are also diamond fields in the Chinese province of Shantung, in Thailand, and on the island of Borneo.

Table 2.2 – Natural mined diamond production, 1962–1982 [215], m ct

Country \ Year	1962	1968	1976	1980	1982
Zaire	16.4	17.5	17.0	14.0	12.2
USSR	2.5[1]	8.5[1]	12.0[1]	12.0[1]	12.0[1]
Republic of South Africa	3.3	7.4	7.3	8.7	8.9
Botswana	—	—	2.4	5.1	8.0
Angola	1.1	1.7	0.4	1.5	1.4
Namibia	1.0	1.7	1.7	1.6	1.0
Ghana	2.9	3.0	2.2	1.1	1.0
Venezuela	1.1	0.1	0.6	0.8	1.0
Brazil	0.3	0.3	0.5	0.5	0.5
Liberia	0.9	0.7	0.5	0.5	0.5
Tanzania	0.6	0.7	0.5	0.3	0.4
Sierra Leone	1.8	1.7	1.1	0.6	0.4
Central African Republic	0.3	0.6	0.4	0.3	0.3
Ivory Coast	0.2	0.2	0.2	0.1	0.1

[1] Estimated.

Note: Since the mid-1980s Australia has risen to dominance, producing about 35 m ct in 1990.

At the present time Africa (Fig. 2.6) and Australia are the greatest suppliers of natural diamonds. In the Republic of South Africa there are a number of diamond mines including the Premier mine in the Transvaal, as well as the Wesselton, Bultfontein, Jagersfontein, Postmastburg, Barkly West and Rustenburg deposits. These sources supply a large quantity of gem quality diamonds and many of the famous large stones were found in this region (Tables 8.2 and 8.3). Because of the high quality and per unit value of single stones, these sources continue to be operated even though they are not very rich in diamond, relatively speaking, and the mining costs are high. For example, the Koffiefontein mine in the Jagersfontein region has an average grade of less than 0.1 ct per 1 tonne of rock. Diamonds are also recovered from secondary deposits along the Namibia coast and from the meanders and terraces along the River Vaal and its basin, e.g. Lichtenburg. Secondary deposits occur also in various parts of the Orange River basin [27].

A greater part of Namibian diamond production derives from rock

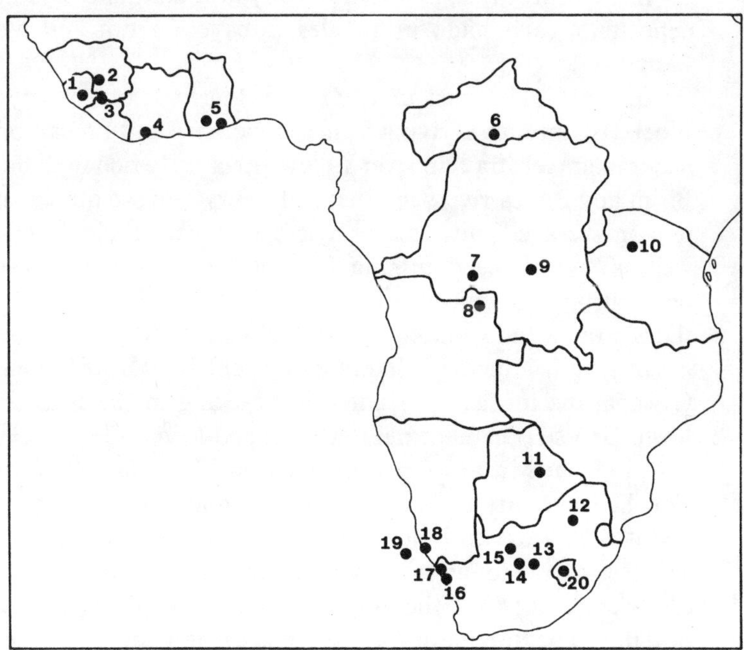

Fig. 2.6 — Diamond producing areas in Africa [214]: *1* — Sierra Leone, *2* — Guinea, *3* — Liberia, *4* — Ivory Coast, *5* — Ghana, *6* — Central African Republic, *7* — Tshikapa–Zaire, *8* — Angola, *9* — Bakwanga–Zaire, *10* — Mwadui–Tanzania, *11* — Orapa–Botswana, *12* — Premier Mine, *13* — Koffiefontein, *14* — Kimberley, *15* — Finsh Mine, *16* — Annex Kleinzee–Namaqualand, *17* — State Alluvial Diggings, Alexander Bay, *18* — C. D. M., Namibia, *19* — Marine Diamonds, Namibia, *20* — Lesotho.

material deposited by the Orange River, and from the coastal sandstone and gravel terraces in the area south of Luderitz. Deposits on the continental shelf are also worked. It is estimated that the coastal and undersea deposits in the region have so far yielded some 30 m ct of diamond, 90–95% of gem quality. The crystals are predominantly colourless or with a light shade of brown [175].

Diamond mining in Botswana was started in the 1970's in the Orapa, and later Lethlakane, kimberlite pipes. The deposits are relatively rich in diamond but only a small proportion is of gem quality. Similarly to Zaire and Australia, the mines yield mostly industrial diamonds. A characteristic feature of diamonds mined in Zaire is their opaque 'milky' structure and their colour, ranging from white-grey, via yellowish to olive and greenish. This kind of diamond is commonly known as 'Congo'. The main diamond mining regions are Bakwanga and Tshikapa in Zaire, while kimberlite pipes are also worked in Tanzania, south of Lake Victoria, and in Angola (Comute, Catoka).

In Central Africa, in addition to typical diamond crystals, there are deposits of carbonado in Tal des Cobaye, Carnot and Nola, west of Bangui.

In the Gold Coast area (Ghana, Sierra Leone, Ivory Coast and Liberia) diamonds deriving from sedimentary and Precambrian magma occur in gravel strata in river valleys. Relatively rich in diamond are the Birim and Bonsa rivers in Ghana. The crystals are quite small, about 1 mm on average. Quite common here are green and brown diamonds, as well as 'bort'. The dominant form is the crystal twin, especially the octahedron and dodecahedron. Larger and more beautiful perhaps are the Sierra Leone stones.

In South America, diamond is mined in Brazil, Venezuela and Guyana, the most important sources being in the Brazilian states of Mato Grosso (Diamantina, Begagen) and Minas Gerais (Grao-Mogol), and along the Mazaruni, Paragua, Cuyuni, Ventuari and Orinoco rivers. For the most part, South American diamonds are of very high quality: about 40% are used in jewellery and 30% are high quality industrials. The crystals are found in conglomerates and breccia in river beds and in old river terraces. Another important diamond area is the Brazilian state of Bahia. The dominant form there is carbonado which, depending on the deposit, accounts for 10 to 70% of the diamond produced.

Extensive diamond mining also takes place in Australia, where the scale of the operation has increased significantly in recent years and is still on the increase. The Argyle mine in the north-west of Australia is currently the biggest producer of diamond, by volume, in the world today.

In contrast with the rest of the world, the main Soviet diamond fields are in exceptionally disadvantageous climatic and terrain areas: swamps, extremely low temperatures in winter, taiga. The first house in thc region for diamond miners was erected in 1957, following which the town of Mirny was built, the 'diamond capital' of the USSR. In addition to the Mirny region, rich diamond deposits have been found in other areas of Siberia, notably the Udachnaya and Aikhal mines (Fig. 2.7). Apart from the Siberian fields, diamonds have been discovered on the Kola peninsula, in the Urals (the earliest USSR source), on the Black Sea coast and in northern Kazakhstan. Siberian diamonds are extracted primarily from kimberlite pipes, and economically recoverable diamonds from secondary deposits are relatively rare. Diamonds found in the USSR are predominantly octahedra and dodecahedra, as well as intermediate forms. In most cases the natural crystals are slightly flattened along one of the axes of symmetry, especially the ternary axis and sometimes, but

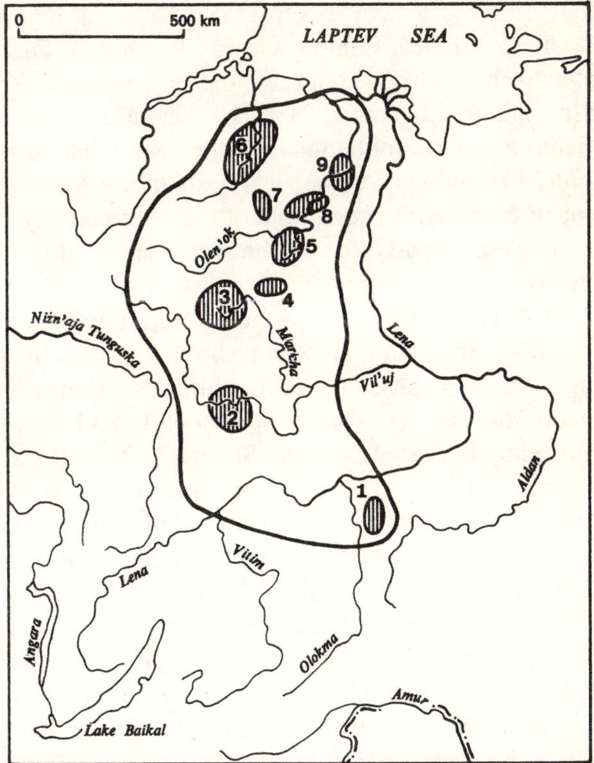

Fig. 2.7 — Major diamond-bearing regions in the USSR [175]: *1* — Aldanskii, *2* — Malo-Botuobinskii, *3* — Daldyno-Alakitskii, *4* — Verchne-Munskii, *5* — Tschomurdanskii, *6* — Anabarskii, *7* — Lutschakan-Kuranachskoi, *8* — Sredne-Olenekskii, *9* — Kuoiksko-Merrschimdenskii.

much less frequently, they are elongated. A considerable proportion are colourless or yellowish.

Kimberlite pipes also occur in North America (USA and Canada), the small Pike County, Arkansas mine being however the only one of any significance. In Europe, no deposits of practical interest have yet been found. Sporadically reported finds in Spain and (much earlier) in Scotland turned out to be geological curiosities.

2.2 SYNTHETIC DIAMONDS

For centuries nature met the demand for diamond crystals. The desire to master the secret of diamond synthesis goes back to antiquity, and has occupied the minds of countless generations of alchemists and scientists. Work on diamond synthesis has claimed many a fortune, and numerous efforts ended in personal tragedy. Diamonds and diamond manufacture have held the attention of Tennant, Lavoisier, Hannay, Moissan, Marsden, Khrushchev, Nernst, Rossini, Tamman, Jessup, Leipunskii, Doelter, Hershey, Gunter, Gesele, Rebentish, Parsons, and numerous other outstanding engineers and scientists. In 1908 the celebrated French physicochemist H. L. Le Chatelier said, “Work on artificial diamonds is to contemporary chemists what looking for the philosophers’ stone was for the alchemists of the Middle Ages”. Success was not to come until the middle of the 20th century.

During World War II, the American companies General Electric, Norton Company and Carborundum Company entered into an agreement aimed at diamond synthesis. Advantage was taken of the work of Percy W. Bridgman, who won the Nobel prize in 1946 for research in high pressure physics. After countless attempts, the first synthetic diamonds were eventually obtained. The General Electric team was headed by F. P. Bundy, H. M. Strong, H. T. Hall and R. H. Wentorf. The first report of the success was published on 15 February, 1955 in *Nature* [77] and by the end of 1957 GE had mastered the synthesis process to such an extent that synthetic diamond became a commercial commodity.

Several years after the publication of the GE report, the ASEA company of Sweden stated that its staff had been working on diamond synthesis since 1942. The chief experimenters were H. Liander and E. Lundblad, and their team had obtained synthetic diamonds in 1953. In that work, use was made of presses designed by B. von Platen where several hundred kilometres of steel wire were wound around the frame housing the chamber in which the high pressure process was carried out. Such a press design is widely used nowadays for diamond synthesis.

Ever since diamonds were successfully synthesized, the production technology has been continually improved. At the same time, entirely new technologies have been developed and industrial diamond production has been launched in many countries all over the world. At the present time, a number of approaches to diamond synthesis are available, the most important of which are [212]:

— growth in a molten solvent-catalyst,
— shock-wave synthesis,
— direct transformation in the absence of a solvent,
— growth on a seed diamond crystal,
— growth under metastable conditions.

Apart from synthesis involving structural transformation of other carbon materials, a number of technologies have been developed to obtain large natural crystal substitutes via the sintering of fine particles. Technologies have also been devised facilitating modification of the properties of diamond by radiation or temperature (Chapter 1).

2.2.1 High pressure static methods of diamond abrasive production

The transformation of graphite—an inexpensive and readily accessible carbon material—into diamond requires a structure condensation by almost 1.5 times. Bundy [4, 80] has demonstrated that it is only at pressures exceeding 6 GPa that fluid carbon is heavier than graphite, and diamonds are obtained from pure graphite at pressures in excess of 13 GPa and high temperatures (about 3300 K). The diamond obtained under such conditions had a fine crystalline structure. Direct graphite-to-diamond transformation under static high pressure conditions is of

Table. 2.3 – Minimum pressures and temperatures for diamond synthesis in a metal-graphite system [80]

System	Pressure (GPa)	Temperature(°C)
Inconel + graphite	4.5	1150
Mn, Cu (12:1) + graphite	4.8	1400
Co + graphite	5.0	1450
Mn, Co (12:1) + graphite	5.0	1450
Mn, Ni (12:1) + graphite	5.3	1475
Ni + graphite	5.5	1460
Pt, Co (4:1) + graphite	5.5	1500
Fe + graphite	5.7	1475
Mn + graphite	5.7	1500
Ta + graphite	6.5	1800
Pt + graphite	7.0	2000
Cr + graphite	7.0	2100

little practical importance. In practice, diamond synthesis is carried out in a carbon–metal system. The presence of certain metals as catalysts distinctly improves the rate and yield of the transformation and, at the same time, makes it possible to lower the pressure and temperature (Table 2.3).

For the commercial production of synthetic diamond abrasives, the most widely used solvents are metals from the VIII group of the periodic table (Co, Fe, Ni) plus manganese. These metals are used in elemental form or as alloys [80, 148, 212].

Figure 2.8 shows the arrangement employed by CeMat'70 (Poland) where metal and graphite rings are placed alternately one on top of another. The synthesis parameters (pressure, temperature, time) together

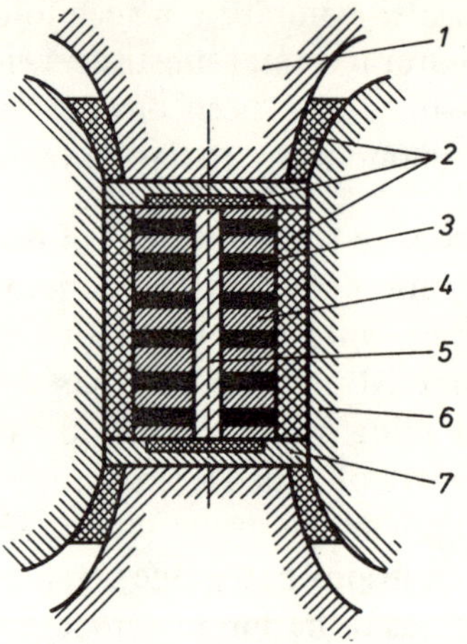

Fig. 2.8 — The charge used in diamond synthesis (CeMat'70–Poland): *1* — sintered carbide punch, *2* — ceramic elements, *3* — graphite ring, *4* — cobalt ring, *5* — steel pin, *6* — sintered carbide die block, *7* — steel ring.

with the composition and arrangement of the charge determine the quality of the product and, by the same token, the economics of the process. The temperatures and pressures typically employed under industrial conditions are 1350–2100 K and 4.5–10.0 GPa [14, 80, 198, 212]. Diamond is recovered from the post-reaction synthesis mass after the metal and unreacted graphite have been dissolved in acids, and the remainder treated in heavy liquids. The basic steps of the synthesis process are depicted in Fig. 2.9.

The production costs and the sales price of synthetic diamond abrasives depend to a large extent on the capital costs for the construction and operation of the high pressure equipment. So far, the most widely used are the two-piston presses modelled on the original

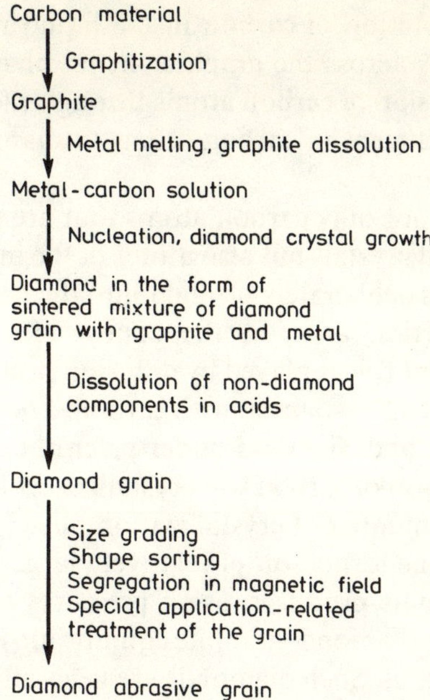

Fig. 2.9 — Basic stages in the synthetic diamond production process.

General Electric and ASEA designs, although four- or six-piston devices are also available. Of utmost importance in diamond synthesis is control of the pressure and temperature, as well as the coaxial action of the pistons. The *temperature* can be measured by determining the parameters of the current flowing across the chamber in which the transformation takes place, or by using a thermocouple designed to work under high pressure conditions [153]. In order to determine the *pressure*, one monitors the oil pressure under the piston of the hydraulic press mated with the chamber. As a rule, the relationship between the pressure inside the reaction charge and the pressure read off the hydraulic press gauges is established experimentally by "calibrating" the apparatus using substances which undergo polymorphic transformations at known pressures. The point at which transformation occurs is recorded, based on the resistance level of the current flowing through the charge.

Depending on the free energy levels between diamond and graphite in the reaction mix, in the course of nucleation and crystal growth diamonds with different structures—and different properties—are formed. The metals used in synthesis play the role of carbon solvents and, at the same time, they are a kind of phase transformation catalyst.

Three stages can be distinguished in the mechanism of diamond nuclei growth:

1. Dissolution of carbon in the liquid metal, i.e. transition of carbon atoms across the graphite–metal phase boundary;
2. Diffusion of carbon atoms through the metal to the growing crystal;
3. Attachment of carbon atoms to diamond edges and faces.

It is not only carbon atoms that are incorporated into the growing diamond crystals but also atoms of the metals used for synthesis, as well as atoms deliberately brought into the process or incidentally present in the reaction area. In the course of crystal growth, a particularly important role is played by diffusion processes and also supersaturation of the metallic solution with carbon. The differences in the solubilities of graphite and diamond under given thermodynamic conditions decide which form of carbon will crystallize. At the same time, the presence of a smaller number of crystallization nuclei and a longer growth time will favour the formation of relatively larger crystals.

Crystals produced using pressures much higher than the thermodynamic diamond–graphite equilibrium line are aggregates of numerous fine crystals. Such diamonds are generally extremely friable, and exhibit various kinds of structural defects. The closer the conditions of the process are to the equilibrium line, the smaller the number of fine crystals and aggregates. If there is enough time during synthesis, well-developed flat-faced diamond crystals are created, which are more regular in shape than the usual run of natural diamonds. As a rule, the slower the growth the better the quality of the crystals [27, 60, 77, 80].

The most characteristic crystal shape growing under the lowest possible temperatures and pressures is the cube. Crystal growth usually takes place in the form of square layers parallel to the face. The layers begin growing in the middle of the face which remains smooth even though it is made up of a series of fine steps overlapping towards the elevated middle. At higher synthesis temperatures, the characteristic octahedral faces become increasingly more prominent (Fig. 2.10). An intermediate shape of synthetic diamond is the cubo-octahedron. Sometimes spiral or layered growth is observed on the $\{100\}$ face of synthetic diamonds. In the case of the $\{110\}$ face, the growth layers are often triangular. Such a pattern arises if, at high synthesis temperature, local dissolution of the crystal occurs. The faces may then become partially etched or develop pits. According to Tolansky [201], trigons are very common on natural diamonds and quite infrequent on synthetic diamonds. With synthetic octahedra one also occasionally observes trigons with opposite orientation. The $\{111\}$ faces are much more resistant to this kind of etching than the $\{100\}$ and $\{110\}$ faces. Crystals of rhombic dodecahedral habit are rare among synthetic diamonds while twins are very common, especially of octahedra and cubo-octahedra. Disintegration of aggregate crystals may yield several diamonds. The faces of the resulting particles are then partially chipped and partially flat [10].

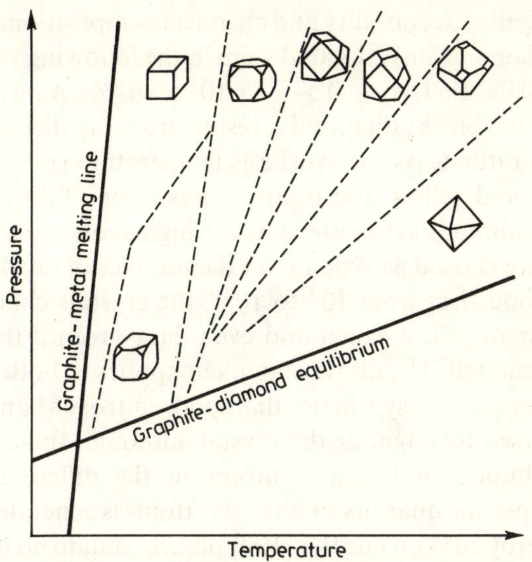

Fig. 2.10 — Synthetic diamond habit vs. pressure–temperature parameters of the synthesis process [18].

Impurity inclusions are very common in synthetic diamonds, especially if the growth was fast. Under conditions of slower growth, the inclusions are more likely to assume positions related to the directions of symmetry in the crystal and the lattice. The inclusions are particles of the solvent metals and other materials involved in the synthesis and they may account for up to 10 or so per cent of the diamond mass. Inclusions may take the form of metal, metal carbide or ceramic clusters, or atoms 'built into' the diamond lattice, substituting for carbon atoms [3, 27, 60, 80, 148].

The inclusions may or may not be oriented relative to the crystal lattice and their presence has a marked effect on the magnetic, optical, mechanical and thermal properties of diamond. It has been found that the {111} planes of metal in diamond are parallel to the {111} planes of diamond; the divergence does not exceed 12° [32]. The volume of metallic inclusions decreases with increasing synthesis temperature. Ceramic inclusions in synthetic diamonds derive from the gasket(s) sealing the capsule. The amounts of aluminium, silicon, magnesium and calcium in synthetic diamond particles grown in the Ni–Mn–C system does not exceed 10^{-2} wt % [22]. Non-diamond carbon may be present in the form of graphite or carbides; for example, crystal phases of Ni_4C and NiC have been identified in diamond in amounts of 0.2 and 0.4 wt %, respectively [165]. Studies of diamonds synthesized in the Ni–Mn–C system, carried out in the USSR, revealed a correlation between magnetic susceptibility and chemical composition. The metal contents in the diamonds investigated were in the following ranges: Mn—0.2–2.33; Ni—0.09–1.69; Fe—0.5–5.4×10^{-2} wt %. As a rule, non-translucent grey or black diamonds result from syntheses conducted at low temperatures. As the synthesis temperature rises, the colour changes to green and yellow and becomes less strong [148].

Diamonds synthesized under high static pressures contain nitrogen and are classed as type Ia. As the number of nitrogen atoms per cm^3 of diamond rises from 10^7 to 10^{20}, the crystals change their colour from yellow to yellow-green and even dark green if the nitrogen content is high enough. The above statement applies to both synthetic and natural diamonds. In synthetic diamond, nitrogen and boron atoms are dispersed throughout the crystal, although there are differences in the distribution of impurity atoms in the different crystal planes. For example, the quantity of nitrogen atoms is generally greatest in the $\{113\}$ plane (by 20%), while the $\{100\}$ planes contain no boron. The presence of nitrogen in natural diamonds reduces their thermal conductivity, while in synthetic diamonds this effect is considerably smaller [148]. Using suitable ultrahigh pressure and temperature treatment, single N atoms can be made to aggregate and N platelets to dissociate.

2.2.2 Static high pressure synthesis of gem quality diamonds

The methods of diamond production described in Sections 2.2.1 and 2.2.3 do not yield diamonds suitable for use as gemstones. The main drawbacks of the diamonds thus obtained are their small (i.e. < 1.0 mm) dimensions — too small to be of any use in jewellery—and various kinds of defects of their internal structure. Methods of synthesizing gem quality diamonds were not patented until 1967, when R. Wentorf of General Electric Co. [77] succeeded in carrying out controlled crystallization of large diamonds grown on seed crystals. The growth took place in the pressure and temperature region where diamond is stable. The pressures employed were about 6 GPa and the presses were similar to those used in diamond abrasive production. However, the capsule in which the process took place had a highly specific design (Fig. 2.11): small diamond crystals were placed in the central region of the capsule in contact with the metal (e.g. iron, nickel) which melted at high temperature and caused dissolution of the carbon present in it. A characteristic aspect of the process was that the central region was heated to a higher temperature (up to about 1730 K) than the top and bottom regions (by about 30 K). Consequently, more of the carbon was dissolved in the central region than in the cooler parts. The carbon atoms diffused to the cooler region where they were 'precipitated' from the liquid metal and

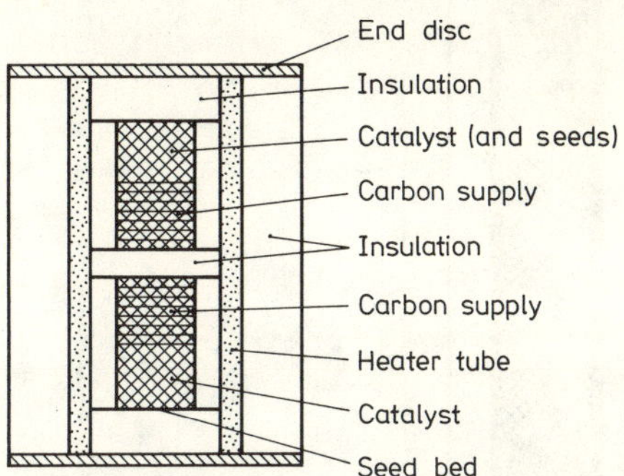

Fig. 2.11 — Arrangement used for the growth of large diamonds using transport of carbon in a metal solution under pressure [77].

deposited onto the seed crystals. The growth rate depended on seed crystal size and decreased as the crystals grew larger: for crystals of about 1 mm in diameter, it was almost twice as high as for 5 mm crystals. A gem size crystal took several days to grow, the average growth rate being 2–3 mg/hr. The best crystals obtained in this way are comparable with the best natural stones, and their weight may exceed one carat.

2.2.3 Dynamic methods of diamond synthesis

A specific way to obtain very high pressures is to harness an explosive shock wave controlled for direction. The pressures that can thus be obtained are several times higher than by known static methods. The possibility of creating diamonds by the use of explosives was considered relatively early, but work that led to successful synthesis was not undertaken until the late 1950's. In 1959 J. Van Tilburg was granted a patent for a process of diamond making with the use of explosives and, at about the same time (1961) and independently of Van Tilburg's work, P.S. De Carli and J.C. Jamieson described the results of their experiments with direct graphite-to-diamond transformation with the use of an explosive shock wave. In 1968 a team working for Du Pont de Nemours and Co. (USA) patented the shock wave synthesis of diamonds [77].

Du Pont is today the largest producer of shock formed diamonds, so-called 'cannon balls' that are very small in size. The set-up employed and the mode of operation are schematically depicted in Fig. 2.12. The

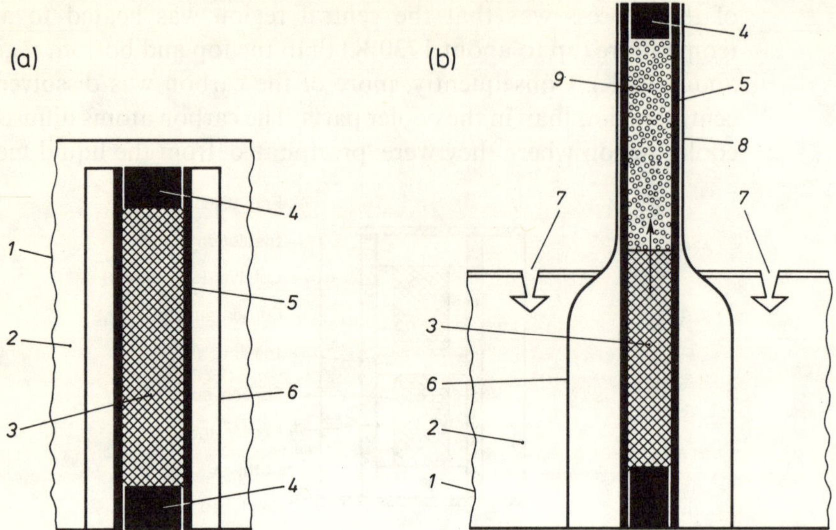

Fig. 2.12 — Arrangement for dynamic diamond synthesis by the Du Pont method [20, 73]: (a) before explosion; (b) after explosion: *1* — screening-drainage tubing, *2* — explosive, *3* — graphite, *4* — stopper, *5* — production pipe, *6* — conveying pipe, *7* — ring explosion front, *8* — deformed conveying pipe on the production pipe, *9* — area in which diamond has formed.

production unit basically consists of a set of concentric tubes. The internal (so-called production) tube is filled with graphite and powdered metal. The production tube is placed inside a driver tube. The diameter of the driver tube is much larger than that of the production tube and it is made of thick steel. The surrounding culvert is the container for several tonnes of a special Du Pont explosive. This assembly is loaded and detonated in a large underground chamber. The explosion takes place at the top of the assembly. A ring-shaped shock wave travels downwards compressing the production tube. This generates a very high pressure in the production tube, of the order of 15–50 GPa, which deforms the tube causing it to become elongated towards the top. The graphite-to-diamond conversion occurs in microseconds. Under these conditions, there is no time for the processes of crystal growth which produce the monocrystalline structures found in natural or most statically synthesized diamond. The time-under-pressure is sufficient for the diamond microcrystals to grow to approximately 10 nm. The Du Pont standard grades have particles which contain from several hundred to many thousands of microcrystallites each, depending on particle size. These microcrystallites are bonded together in a random polycrystalline structure without significant cleavage planes. Therefore, the particles are equally strong in all directions. X-ray spectra for different kinds of diamond are compared in

Fig. 2.13. The diffraction peaks show considerable broadening, which is characteristic of polycrystalline materials having a small crystallite size. A standard method of BET nitrogen surface area measurement shows that 6 μm particles of Du Pont diamond have 300% of the surface area of 6 μm natural diamond and 200% of the surface area of 6 μm monocrystalline synthetic diamond [20, 73]. These large surface areas are due to the polycrystalline structure of the diamond. The density of Du Pont diamond is 3.451 g/cm^3. A characteristic feature of this type of diamond is the presence of lonsdaleite (cubic diamond). In terms of their structure, the Du Pont diamonds are closest to natural carbonado. The commercial product is currently available from 1/8 μm to 100 μm in size.

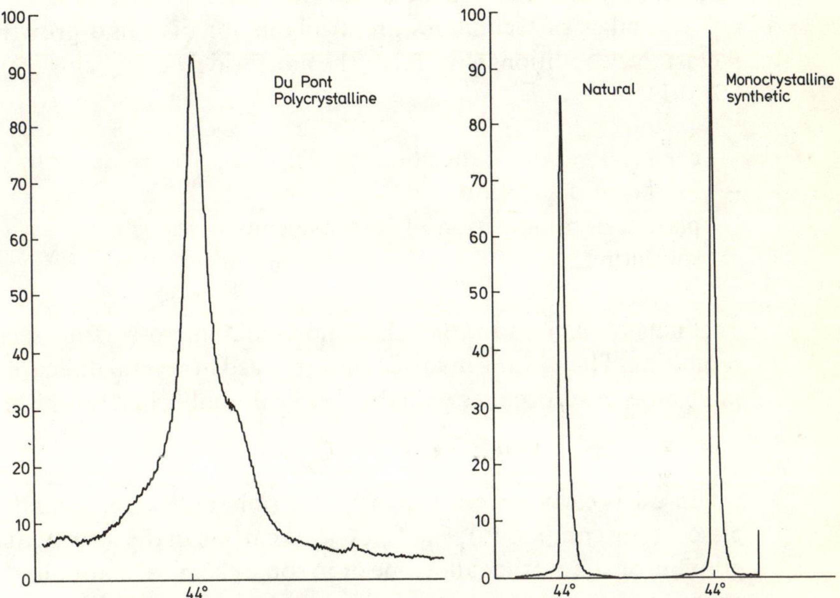

Fig. 2.13 — X-ray diffraction analysis of Du Pont diamonds shows the broadening typical of polycrystalline materials having a small crystallite size [73].

Diamonds are also synthesized by dynamic methods in the USSR, where they are designated AV (for almazy vzryvnyje or explosive diamonds). The pressures employed are in excess of 40 GPa and the pressure build-up is 10–100 GPa/ms [15, 19, 133].

2.2.4 Diamond synthesis under metastable conditions

Transition of carbon atoms from a non-diamond to a diamond lattice involves a change in the internal energy of the system. Direct graphite-to-diamond conversion causes an increase in the internal energy of the system and requires that the carbon atoms be excited to a high energy. In the methods of diamond synthesis described above, this was implemented by a change in the thermodynamic conditions of the system resulting from the application of ultra-high pressures and temperatures. An alternative way to obtain a diamond crystal lattice arrangement of the carbon atoms is to use a carbon material in which the energy of arrangement of carbon atoms (ions or radicals) is relatively higher than the internal energy of the atoms arranged in the diamond structure. In other words, at pressures in the region where diamond is metastable, atoms (ions or radicals) may arise from some carbon compounds in the course of their chemical reaction or dissociation, which may constitute diamond seed crystals or they may be deposited onto seed crystals and augment the diamond structure already there.

A number of techniques are available for diamond growth under metastable conditions (Fig. 2.10), the most widely used being [60, 61, 70, 80, 101]:

— chemical vapour deposition (CVD),
— ion beam deposition,
— plasma decomposition of carbonaceous gases,
— sputtering.

Films of diamond rather than individual micro-crystals seem most promising. The process involved may proceed in several directions at the same time, in accordance with the chemical equilibrium for the reaction:

$$(a+b+c)C_W \rightleftharpoons aC_D + bC_G + cC_X,$$

where C_W is carbon from the gas phase (ion, radical, or atom from the dissociating compound); C_D, C_G, C_X are carbon in the form of diamond, graphite or some other allotrope or in some chemical compound; *a*, *b*, *c* are proportionality coefficients. The conditions under which the process is conducted and the character of the substrate should be such as to permit "termination" of the crystallization at the point when the maximum quantity of diamond has formed.

Epitaxial methods of diamond growth make it possible to control the structure of the growing micro-crystals and thus to obtain diamond products with properties not found in nature. It is relatively easy to introduce into the crystal lattice—in the place of, or side by side with, the carbon atoms—atoms of elements near to carbon in the periodic table, i.e. boron, nitrogen, aluminium, phosphorus, etc. The results of epitaxial growth depend on a number of factors, the most important of which are the composition and concentration of the gas which contains carbon among its components, the gas flow rate, temperature, design of the equipment, especially gas supply and heating. Particularly well described in the literature are systems involving methane with other hydrocarbons or halogen derivatives, and methane with hydrogen [101, 188]. Good results are obtained by heating the gas to 1000–1300 K with the overall gas pressure at 10 to 100 Torr. A crystal a few μm in size takes several hours to grow [60]. Higher methane pressures favour the formation of spherical micrograins, while those obtained at lower gas pressures are flat-walled with the structure of compact isometric blocks. The characteristic faces are $\{111\}$ and $\{100\}$, with the ratio of the surface areas of these faces depending on the methane content. At higher methane concentrations and at higher temperatures, graphite forms. Good results can also be obtained using hydrocarbon mixtures with hydrogen at 950–1900 K. The use of hydrogen makes it easier to remove the non-diamond carbon, removal usually being carried out during the second stage using hydrogen at a pressure of several hundred atmospheres and at a temperature of about 1300 K [101].

Derjagin *et al.* [60, 61] experimented with the growth of diamond crystals in the form of fibres or filaments, needles and plates using epitaxial methods. The process was conducted in methane at pressures below atmospheric. It was found that diamond growth occurred first of all on the $\{111\}$ faces. The growth rate was 10 μm/h in the case of a face size of 25 μm.

Some other methods of production in the region where diamond is metastable make it possible to obtain thin carbon films with a diamond-like structure. In the literature, carbon films are known as a–C:H (amorphous hydrogenated carbon) films, i–C (ion carbon) films, diamond-like films or diamond films [70, 101]. In the ion beam techniques, carbon films are obtained from carbon ions accelerated in a constant electric field. The RF (radio frequency) plasma method differs from the first one in that ions are obtained and a constant electric field is generated in which these ions are accelerated. Ions, produced as a result of hydrocarbon decomposition in a glow space, are accelerated in the electric field created between the positively charged plasma of the glow discharge and the negatively self-biased electrode. Depending on the deposition conditions, films are obtained with different electrical properties. At the same time, such films show high abrasion resistance as well as optical properties (a particularly wide range of electromagnetic transmittance) which are close to those of large monocrystalline diamonds.

2.3 POLYCRYSTALLINE DIAMOND (PCD) TOOL BLANKS

Relatively large size (up to 70 mm diameter) natural diamond substitutes can be obtained by sintering fine diamond particles together to form a polycrystalline mass. The final product may comprise a monolithic or layered profile with precisely defined stereometry. Such blanks can be used—directly or after appropriate mechanical, wire EDM, or laser beam cutting, for the production of cutting tool, mining bit, wire drawing die or wear resistant part blanks. Commercially marketed blanks have linear dimensions of up to about 50 mm. In the USSR, the comminution of PCD products yields coarse mesh abrasives.

An important characteristic of synthetic polycrystalline diamonds, differentiating them from natural and other synthetic diamond crystals, is the isotropy of their physical and chemical properties. In terms of structure, PCD is quite different from mono- or polycrystalline natural or synthetic diamond. The exceptions are those PCD products (e.g. the Soviet-made SVSP and ARS-3) which are similar in structure to carbonado. Synthetic PCD may also be porous. Tough and cohesive PCD materials can only be obtained under ultra-high pressure and temperature conditions. With diamond, plastic strain is observed at temperatures above 1570 K in the region of its thermodynamic stability. It is also under these conditions that particle recrystallization is possible, which is observed to proceed via the intergrowth of fine particles to form coarser ones. Under such conditions even partial graphitization of the particle surfaces favours displacement of carbon atoms and the formation of the diamond phase at inter-grain boundaries. Also advantageous is the use of diamonds with rough surfaces, e.g. diamonds obtained by dynamic synthesis or diamonds with partly etched faces [60]. The addition of a substance which acts as a kind of binder facilitates diamond intergrowth but, at the same time, it changes the structure of the blank and modifies its properties. The 'binder' phase may be carbide or metal, while boride and nitride phases are quite infrequently employed. The 'binder' may be introduced in the form of a powder or as a coat on the diamonds. It may also hold the diamonds together as a result of liquid metal infiltration, e.g. cobalt from a carbide matrix.

In the course of high pressure sintering, the diamonds may react with the bond leading to the formation of intermediate layers at the diamond-bond interfaces. The thermal and mechanical properties of these intermediate layers make the phase boundary between diamond and the bond less abrupt. This is an advantageous development, for abrupt phase boundaries usually generate various kinds of thermomechanical stresses owing to differences in the thermal expansion of the different components and in their resistance to mechanical stimuli. In Fig. 2.14a the thermal expansion coefficients of the substances used by De Beers for the production of PCD blanks are shown.

In Fig. 2.14b, on the other hand, the abrasion rates of PCD materials known under the trade names of SYNDITE and SYNDAX-3 are compared. The latter material is obtained by ultra-high pressure sintering of diamonds 15–30 μm in size with silicon powder [194]. Sintering leads to the formation of β-SiC which, at the same time, helps bind the diamonds together. The silicon carbide content varies between 16.5 and 18.9 vol %. The excellent mechanical strength and thermal resistance of diamond bonding is in this case probably related to the similarity of crystal structure and bond types in diamond and β-SiC. SYNDITE, on the other hand, has a cobalt matrix and is produced at a temperature of about 1700 K and a pressure of about 6 GPa, in which the cobalt content may be as high as about 8 vol % [1]. The application fields of the two PCD products are defined by the differences in the expansion coefficients of the binders employed in making them. The use of ferrous metals is conducive to the formation of carbides and graphitization when the blank is being heated. At the same time, a large difference in the expansion coefficient of the binder relative to diamond

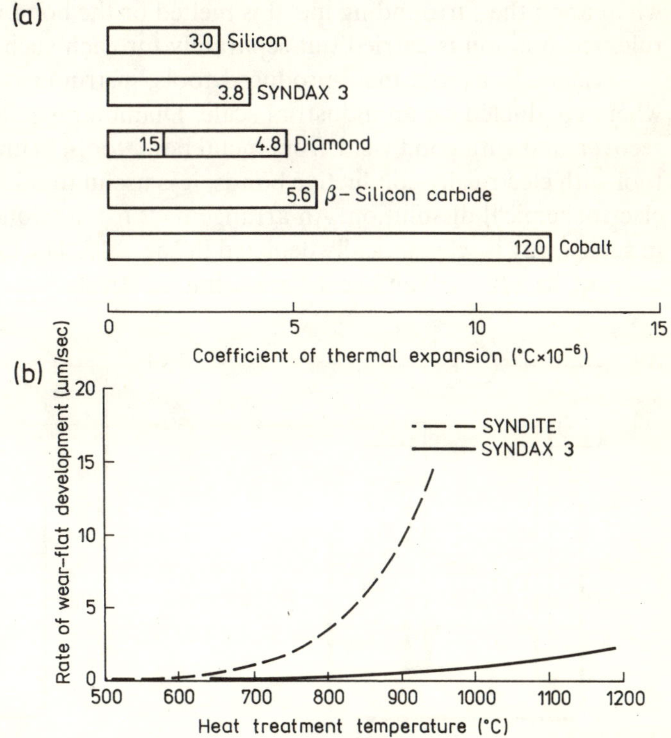

Fig. 2.14 — (a) Thermal expansion coefficients of the components used in the manufacture of PCD materials by De Beers [204]; (b) rate of wear-flat development of two different diamond 'sinters' in the course of turning of granite as a function of temperature [204].

increases the risk of the blank fracturing at above about 1000 K. One way to improve the thermal resistance of synthetic PCD is by elimination of the cobalt binder, e.g. by leaching it chemically. In such a case the blank acquires a porous structure. Another way is to avoid the use of a metal binder altogether, as in the case of De Beers' SYNDAX-3.

2.4 DIAMOND RECLAMATION FROM WORN TOOLS

The relatively high price even of impure industrial diamond and the growing demand for this raw material makes recycling of broken or worn tools a source of diamond one can hardly afford to ignore. In a sense, diamond is used in a kind of 'closed loop'.

The procedure to follow depends on the tool or product type and on the potential value of the diamond contained in it. Where crystals of high value are involved, as in jewellery, diamond elements in research instruments, electronics parts, or the so-called single crystal tools, diamond reclamation consists in mechanical disassembly of the product whereupon the surrounding metal is melted or the bond is etched. As a rule, reclamation is carried out separately for each such product.

Reclamation from mass-produced tools, instruments, etc. is everywhere conducted on an industrial scale. Diamond is relatively easy to recover, and with good yield. from metal bond tools. With most types of tool with electrically conductive bonds, it is useful to take advantage of electrochemical dissolution. An arrangement for diamond reclamation in such a way is schematically depicted in Fig. 2.15. The electrochemical method has the advantage of releasing relatively small amounts of

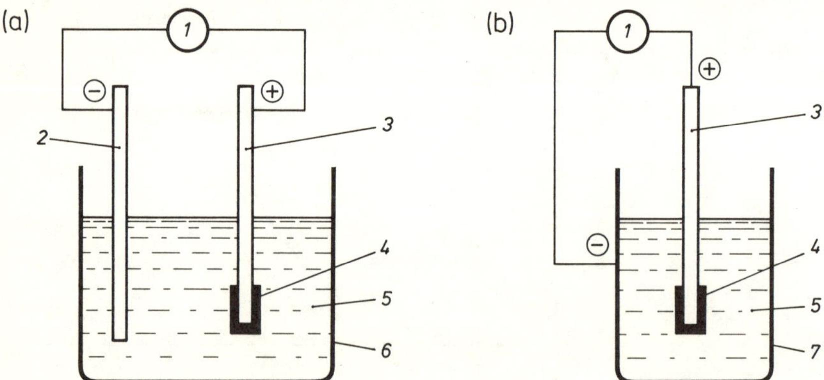

Fig. 2.15 — Electrical circuits for diamond reclamation by electrochemical decomposition of tools: *1* — D. C. supply, *2* — auxiliary electrode, *3* — metallic tool blank, *4* — diamond-containing part of the tool, *5* — galvanic bath (electrolyte), *6* — tank, *7* — electrode in the form of tank.

Table 2.4 – Diamond reclamation from worn tools [13]

Type of tool	Recommended method	Reclamation product and reconditioning possibilities
Single-point tools	De-brazing the diamond Dissolving the matrix in acids Electrolytic dissolution	Diamonds directly applicable in manufacture of new tools. Visually examined to determine the state of fracture and extent of cutting edge and point wear. Material may also be subject to mechanical treatment, e.g. formation of new cutting points by grinding, ovalization, etc., depending on quality
Multi-point tools with diamond stones	Electrolytic dissolution Dissolving the matrix in acids	
Tools with metal bonded diamond grit sintered at above about 900°C	Dissolving the matrix in acids	The diamond abrasive and so-called fillers (e.g. tungsten carbide, silicon carbide, etc.) may be directly re-usable in the manufacture of new tools
Tools with soft metal bonds	Dissolving the matrix in acids	Graphitization, fractures, and oxidation of diamond surfaces should be checked, together with shape and size. Diamonds can be separated from fillers by mechanical or chemical procedures
Tools with electroplated diamond grit	Dissolving the matrix in acids	
Vitrified bond tools	Etching in hydrofluoric acid Dissolving with alkalis	
Resin bond tools	Oxidation or thermal dissociation of the bond Exposure to solvents	

harmful vapours and waste to the environment. Relatively the least profitable, in economic terms, is the recovery of diamond from resin bond tools, especially grinding wheels (Table 2.4).

The cost of reclamation depends first of all on the labour and energy costs, as well as on the organization of labour, the quantity of tools involved, and the capital costs of the reclamation plant. On the other hand, the maximum value of the products reclaimed and, by the same token, the profitability of reclamation, can be assessed from preliminary examination of a tool and determination of its wear. In most cases, the value of the other reclaimed products is equal to the value of the diamond recovered. The prices of diamonds used in tools are compared in Fig. 2.16.

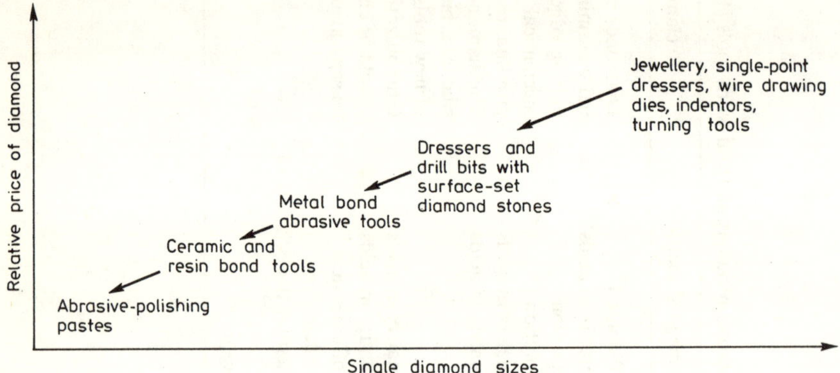

Fig. 2.16 — Relative prices and utilization possibilities of the diamond reclaimed from used tools [13].

It has to be kept in mind, however, that the diamond reclaimed from a tool may be different from the diamond which originally went into it, a circumstance which may impose limitations on its subsequent use. Two kinds of damage may occur to the diamond:

— on particle surfaces, as a result of chemical reaction with the bond, with the work-piece material, or with the coolant, or as a result of particle wear,
— inside the particle as a result of fracture caused by the various thermal-mechanical forces to which it is exposed, or as a result of graphitization.

Obvious changes of diamond particle quality observable with the naked eye or under a microscope include fractures, chipping under the influence of relatively weak forces, etch figures on the surfaces, surface and internal graphitization (black surfaces and internal patches), etc. [13]. As a rule, the material reclaimed from a tool requires additional treatment. The kind of treatment depends on the state of the reclaimed material and on the use(s) to which it will be put.

3

Diamond types and their applications

Even the oldest historical records of diamond refer to two different ways in which the mineral was used, viz. as a gem and in non-gem applications. Gem diamond belongs in the sphere of 'the spirit, beauty and affluence', as used in cult objects, jewellery, or as readily negotiable 'hard currency'. Industrial diamond, on the other hand, is incorporated into multifarious tools and instruments used by man in his work or for investigating the world around him. The division into gem and non-gem or industrial diamonds is still with us today. Only 10–20% of all mined natural diamond is of gem quality and can therefore be used in jewellery [175]: the 'per unit' prices of gem diamonds are very high. Almost all of the non-gem quality natural diamonds, all synthetic diamonds, and all polycrystalline diamond blanks constitute the raw material for technical applications. The approximate breakdown in uses of industrial diamond by size of the crystals or fine particles is shown in Fig. 3.1. It is estimated [80] that some 95% of industrial diamond is used in the manufacture of all manner of tools — the remainder is presumably used in scientific instruments and non-cutting applications.

The division into gem and industrial (natural) diamonds is not a clearcut one and the dividing line is drawn at different places at different times. Considerations related to economic necessity and current levels of supply and demand have made or may make it necessary to use gem quality crystals as industrials, e.g. as in Japan during World War II. The reverse is also true, but to a limited extent: only the highest quality industrial diamonds can be used for making diamond jewellery.

Another traditional classification of diamond is based on optical properties (Table 1.3). Such a four-way division is of limited practical use, because no account has been taken of many of the essential functional characteristics of diamonds. An extremely wide range of uses are made of diamonds nowadays and this may apply to any particular crystal. The first stage in a practical classification of diamonds involves evaluation of a set of characteristics related to the crystal structure, with some indication of the economically viable range of applications. The essential characteristics include: crystal size, shape, and colour, internal purity, and state of the facets. In terms of size, diamonds may be roughly divided into three groups:

— natural and synthetic gem-quality stones, and synthetic PCD tool blanks,
— natural and synthetic diamond abrasives,
— natural and synthetic micron diamond powders.

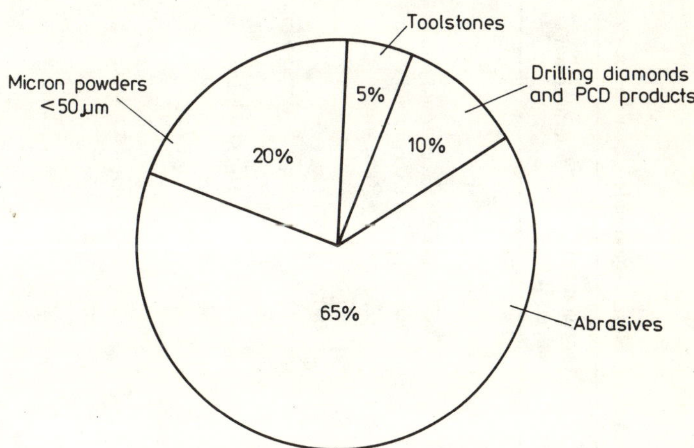

Fig. 3.1 — Approximate consumption of natural and synthetic industrial diamonds [175].

Below we will discuss the different kinds of diamond material in the order listed above. It should be noted, however, that the above classification is an arbitrary one. The linear dimensions of diamonds in each of the three groups vary over a fairly wide range. Each group can be further sub-divided and classified in terms of physico-chemical properties and, by the same token, functional characteristics, etc.

Table 3.1 – Classification of gem diamonds based on their colour [27]

Colour designation	Approximate value (%)	International colour grading	GIA	
1. Jager (blue-white, high blue fluorescence stones from Jagersfontein Mine)	—	Exceptional white +	D	Colourless
2. River	100	Exceptional white	E	
3. Top Wesselton	95	Rare white	G	Colourless through the crown side of the diamond
4. Wesselton	90	White	H	
5. Top Crystal	80		I	
6. Crystal	70	Slightly tinted white	J	Slightly coloured
7. Top Silver Cape	—		K	
8. Silver Cape	—	Tinted white	L	
9. Cape	40		N	Lightly coloured to coloured
10. Very Light Yellow	—			
11. Very Light Brown	—			
12. Light Yellow	—	Tinted colour		
13. Light Brown	—			
14. Yellow	—		S–Z	
15. Brown	—			
16. Dark Yellow	—			
17. Dark Brown	—			
18. Very Dark Yellow	—			
19. Very Dark Brown	—			

3.1 DIAMOND STONES AND SYNTHETIC PCD BLANKS

3.1.1 Diamonds for jewellery

For perhaps two thousand years, flat facetted diamonds have been the highest rated gemstones for jewellery. It has become the norm to say that the value of a polished gem diamond is determined by 'the four C's', viz. "**C**arat (weight), **C**olour, **C**larity, and **C**ut".

There is no straightforward relationship between the weight, crystal perfection and price of a rough (uncut) diamond. Roughly speaking, the value increases exponentially as a function of the diamond weight. However, only a relatively small percentage of stones can be classified as completely colourless. Most natural diamonds have a colour of varying hue and intensity. Of utmost importance in grading the colour of a diamond is proper light. Care should also be taken to avoid ultraviolet radiation, which may affect judgment due to fluorescence. Colour should be examined in white light and against a white, matt background. For grading to be completely free from environmental influences, use is made of special equipment developed by gemmological institutes, e.g. the Diamondlite or Koloriscop. The scale traditionally employed in natural diamond colour grading is shown in Table 3.1. Other scales are also in use, based on spectrophotometric or colorimetric measurements. The

Table 3.2 – Designation of fault in gem diamonds [27]

	Designation of fault	Characteristics
Internal	Solid inclusion	Solid bright or dark inclusion of graphite, hematite, garnet, enstatite, zircon, diamond or sometimes quartz
	Bubbles	Gas or liquid bubbles, usually transparent and of ovalized shape
	Clouds	Groups of small bubbles or solid inclusions
	Feathers	Small surface cracks visible as 'bird feathers'
	Butterflies	Inclusions surrounded with small surface cracks visible as 'butterfly wings'
	Cracks	Cracks incompatible with normal diamond crack directions
	Naats	Twins
	Fezels	Shallow white inclusions in twinning planes
External (surface)	Scratches	Scratches on the surface
	Chips, nicks	Crumbles, chips
	Pits, cavities	Cavities, caverns
	Flats	Flats on the girdle
	Naats	Twinning traces on the surface
	Naturals	Fragments of natural crystal faces
	Bearded girdle	Jagged girdle

Table 3.3 – Purity of gem diamonds [27]

Classification			Characteristics	Relative value
UK	Scand. DN	GIA (USA)		
Flawless (clean)	FL (F[1])	FL	Faultless, perfect, pure. Without visible internal faults under 10 × magnification	100
VVS	VVS$_1$[2], VVS$_2$	VVS$_1$,VVS$_2$	Very, very small inclusion, difficult to distinguish under 10 × magnification	90
VS	VS$_1$, VS$_2$	VS$_1$, VS$_2$	Very small inclusions, detectable at 10 × magnification	80
SI	SI$_1$, SI$_2$	SI$_1$, SI$_2$	Small inclusions, difficult to distinguish under a loupe	70
1st piqué	1st piqué	I$_1$ (imperfect)	Inclusions visible to the naked eye, easily distinguishable under a loupe	—
2nd piqué 3rd piqué	2nd piqué	I$_2$	Bigger or numerous small inclusions visible to the naked eye	—
Spotted	3rd piqué	I$_3$	Large and numerous faults visible to the naked eye	—
Heavily spotted	—	—	Inclusions very easily seen with the naked eye	—
Rejection	—	—	Crystals with large faults, not suitable for jewellery	—

[1] Stone must be without internal impurities.
[2] Steps VVS, VS, SJ are used for larger than 0.5 ct stone valuation.

most widely used are the scales adopted by the Gemmological Institute of America (GIA), Scandinavian Diamond Nomenclature (Scan. DN), and the European Gemmological Laboratories (EGL). The most precious diamonds are colourless with a bluish-white tint. That tint cannot, however, be treated as colour because it is caused by fluorescence which is induced by the ultraviolet component of daylight. Diamonds with light, uniform colours, known as 'fancy stones', are extremely rare. Such stones may be more highly valued and fetch higher prices than colourless diamonds. Brown and canary yellow colour specimens are fairly common, whereas pink, red, blue and green are very rare. Such stones owe their colouring to a distored crystal lattice or, less frequently, to atomic or bulk scale inclusions incorporating, e.g. iron, titanium, silicon and aluminium oxides, as well as pyrope, olivine and other minerals. In pink diamonds, however, the colouring is due to the presence of trace amounts of manganese [27, 148]. Very often an apparent change of colour of a gem diamond is caused by the setting or by the colour of stones set together with the diamond.

Any flaw, even the tiniest, inside or on the surface of a crystal, disturbs the passage of light through the crystal, thus lowering its value as a gem. The various kinds of structural imperfections observed in diamond have their own specific names (Table 3.2). The basic test used in grading the purity of a diamond is the visibility of defects under 10 × magnification. Use is also made of specially designed microscopes equipped with different kinds of lighting. One of the most widely used is the Gemolit instrument developed by the Gemmological Institute of America equipped with a zoom lens and several kinds of lighting. The magnification ranges from 10 to 100 ×. Some diamond purity designations are presented in Table 3.3. As a rule, the designations are based on the first letters of the English terms defining the level of impurity. In addition to the type and size of imperfections, evaluation of diamond purity is affected by colour and, if applicable, the setting.

In ancient times the dazzling optical effects created by a polished diamond, due to light dispersion and refraction, were attributed to magic power and possession of a diamond was said to endow the owner with power, authority, health and wealth. For the luminous effects to become fully evident, the stone must first be cut (faceted) and then polished; these properties are obviously related to the fourth C — cut. The history of the craft of diamond cutting can be regarded as part of art history. Some aspects of gem diamond cutting will be discussed in Chapter 8.

3.1.2 Synthetic gem quality diamond

In 1970 the General Electric Company (USA) announced the production of synthetic diamond crystals corresponding in quality to gemstones. The crystals were obtained as a result of slow, controlled growth of a diamond seed placed in a liquid or gaseous phase saturated with carbon atoms [27, 77, 150]. The weight of such a diamond may exceed one carat (Fig. 3.2).

Present-day gem quality synthetic diamonds are structurally as perfectly developed as the highest quality natural gemstones. In order to obtain the desired optical, electrical, thermal and mechanical properties, as well as the desired inner structure and habit, appropriate dopants are introduced during synthesis. 'Gem synthetics' are now also produced by Sumitomo and De Beers (for example: MONODIE 111 and MONODIE 100—De Beers synthetic monocrystalline diamond die blanks).

The properties of 'man-made' diamonds may be varied over a wide range, it being possible to obtain crystals with properties not found in natural stones. The presence of nitrogen makes them green, the colour changing from yellow to green as the nitrogen content increases. A characteristic constituent in blue diamonds is boron, as is the case with

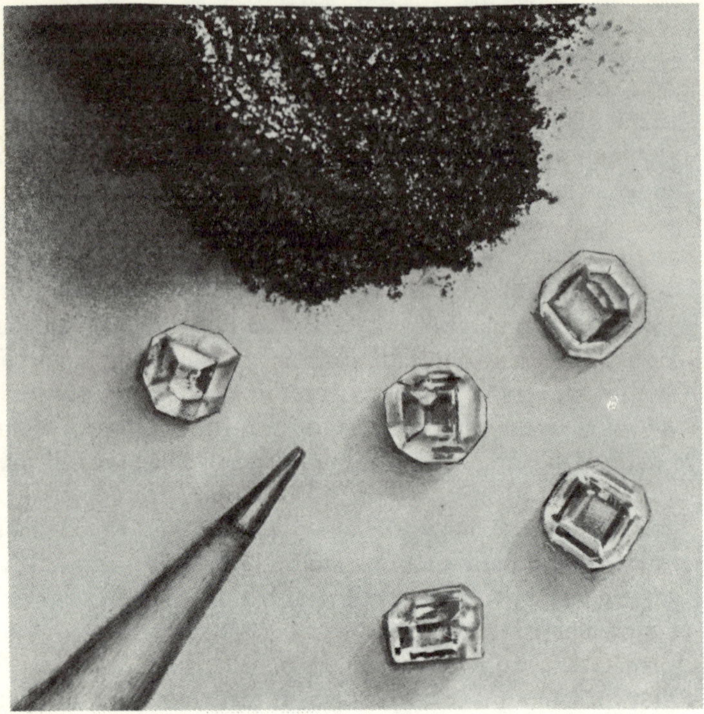

Fig. 3.2 — Gem-grade synthetic diamond crystals obtained at General Electric Co. (USA) laboratories from graphite (the black powder on left). The pencil tip included in the picture gives on idea of the sizes involved (courtesy of General Electric Co.).

natural crystals. Colourless and blue synthetic diamond stones are characterized by relatively high electrical conductivity, strong fluorescence and phosphorescence, with a greater intensity than is generally observed in natural type IIb diamonds. The green-yellow crystals do not conduct electricity. The first diamonds of this type, synthesized by a General Electric team, had tabular habit [167]. Most frequently synthetic and natural stones are distinguished on the basis of quite sophisticated tests, including X-ray and optical tests which involve an analysis of the distribution of impurity elements in the crystal, especially nitrogen.

Thus far synthetic gem quality diamonds have been used on a relatively small scale. This is due to their high price and the fact that they are—synthetic! Nevertheless, the importance of high purity synthetic diamond is steadily growing, especially in electronic engineering where it is used as a semiconducting material and a heat sink. High quality diamond is also useful in different kinds of test devices, being transparent to electromagnetic radiation, chemically inert, resistant to high temperatures, pressures and abrasion, and having a high capacity for heat extraction.

3.1.3 Natural industrial diamond stones

Historically, this kind of diamond raw material is frequently divided into several groups according to the way in which it is utilized, e.g.:

— Toolstones,
— Drilling,
— Diestones,
— Dressers, etc.

The term 'diamond toolstones' has a fairly wide scope and is frequently understood to cover, among others, diamonds used in exploration drilling [198]. The above divisions are thus not clear-cut because some crystals may prove suitable for more than one application.

Natural industrial diamond stones embraces both crystals with a naturally developed habit, and crystals the surface and shape geometry of which were formed by special mechanical-chemical treatment.

3.1.3.1 Diamond toolstones

Diamonds in this group have weights ranging roughly from 200 stones per carat (spc) to stones weighing 5.5 carats. The way in which this kind of rough diamond is utilized depends on the individual characteristics of the stones. Among the class of diamond toolstones one can distinguish those suitable for cutting tools, shaped dressers, plain dressers, wire drawing dies, hardness testing indenters, etc. Individual crystals are rated by weight, habit, and inner and surface impurity content. The characteristic shapes of toolstone type diamonds are shown in Fig. 3.3 and defined in Table 3.4 [71, 198].*

From the point of view of the extent and quality of inner structure development, industrial stones are assigned to two classes:

— near-gem, i.e. high quality crystals approximating the material employed for jewellery. Such diamonds are intended for use in single crystal shaped tools, e.g. measuring instruments, wire drawing dies, dressers with a profiled cutting edge;
— bort (or boart), a material which, in crystallographic terms, is not fully developed and often shows various defects. Such diamonds are used, frequently crushed in multi-grain tools.

---

* A wide range of toolstone type diamonds—all offered by Henri Polak Diamond Corp. (New York, NY USA)—is shown in the Appendix.

Diagrammatic figures of shapes of diamond stones	Definitions
	Octahedron
	Dodecahedron
	Cube
	Round
	Triangular
	Elongated
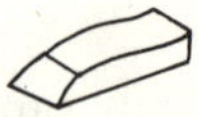	Processed Point
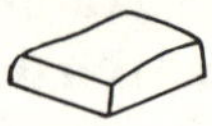	Processed Flat

Fig. 3.3 — Basic development forms of natural industrial grade diamond stones [71].

The conditions under which a given tool is to operate impose specific requirements on the quality of the crystal. Depending on the nature and precision of the machining operation involved, diamonds with different mechanical strength, dimensional tolerance and stereometry are required. Since the stones are exposed to various kinds of mechanical stresses and thermal shocks, they should be so selected as to resist them and only undergo gradual wear due to attrition. In order to eliminate imperfect crystals, the stones are tested under extreme conditions of operation expected for the particular type of tool. For example, they are heated and then cooled abruptly. The highest quality stones, to be used for measuring instruments, wire drawing dies, dressers and cutting tools with profiled edges should show no inclusions under 10 × magnification. There must be no fractures. As a matter of fact, stones with inclusions and microfractures do find use in some operations, e.g. when the machining precision and tool service life requirements are fairly low.

Table 3.4 – Terms of diamond toolstones recommended for general use by ANSI B67.1 — 1978

Shape classi-fication	Geometric classification	Description
1	Octahedron	An eight-sided diamond which can have a maximum of six points
2	Blocky octahedron	An irregular blocky diamond with a basic octahedron form
3	Cube	A six-sided diamond, with parallel sides
4	Dodecahedron	A twelve-sided formation with rounded surfaces. This geometric form has six major and eight minor points
5	Crystal	An octahedron where the sides are flat and come to relatively sharp ribs and points at their intersections
6	Cylindrical elongated	A distortion of the octahedron or dodecahedron basic form, to a generally cylindrical shape
7	Egg shaped	A distortion of the octahedron or dodecahedron basic form to a generally egg shape
8	Flat elongated	A distortion of the octahedron or dodecahedron basic form, to a flat elongated shape
9	Parallel type Macle	A twinned diamond generally triangular in shape and with flat parallel sides
10	Cushion type Macle	A twinned diamond generally triangular in shape with convex sides
11	Ballas	A rare spherical form, cryptocrystalline with an interlaced crystal structure arranged radially around a centre
12	Round	Spherical but with a cubic crystalline structure

Flawed crystals are also employed in the so-called multi-grain tools, where the performance is dictated by a number of stones.

In every case the choice of a stone for a given application should be made on an individual basis, and it should be based on a thorough knowledge of the tool operating parameters. Depending on the requirements imposed by the tool design and the operating parameters, different degrees of crystal perfection may be appropriate. This is due to diamond's hardness and wear anisotropy and, because of this, when making tools it is essential to take account of the particular characteristics of every crystal and its crystallographic orientation. The necessity to take into account the hardness anisotropy forces one to use diamonds where the crystallographic directions are "visible", or else to use X-ray equipment, facilitating very accurate orientation of the crystals [139]. At the same time, exposing the natural cleavage planes and the best directions of wear resistance permits more precise orientation of the stone relative to the workpiece surface. The crystals may be suitable for mounting directly, the way they are, or they may require polishing (grinding) to impart some specific shapes to them. Usually only perfect or near perfect stones are polished, on special machines: flawed crystals might fracture. In terms of habit, the most sought after for toolmaking are octahedra, rhombo-dodecahedra, and, somewhat less popular, cubic crystals.

A good illustration of the necessity to take into account hardness anisotropy is the indenter used for hardness testing, which has to meet high requirements of uniform wear and good mechanical strength. If the indenter symmetry axis is oriented to conform with the quaternary axis of diamond crystal symmetry, these requirements can be satisfied in the most beneficial way. However, if this condition is not met, e.g. if the orientation conforms with the ternary axis, part of the stone may become chipped. Somewhat different are the requirements for diamonds to be used for wire drawing dies (diestones). The bore in these tools should be made in the crystal planes resistant to abrasion, so that they retain the bore profile and dimensions intact for a long time. This can be achieved by orienting the die bore to conform with the ternary axis. Especially useful for cutting tools on the other hand are dodecahedral crystals elongated along the diad and triad (ternary) axes, as well as regular flat-walled octahedra and dodecahedra [139]. Typical diamonds selected weigh from 0.3 to 2 ct. Because of the anisotropy of diamond, the 'hard' and 'soft' directions must be taken into account both in edge forming and in setting the diamond in the tool shank. Similarly, when diamonds are used in single-point dressers the tool performance depends on its orientation relative to the wheel being dressed. Diamonds for single-point dressers are selected for their structural strength, number of points, sharpness, and absence of defects. Proper selection of crystal size and quality for a particular application requires expert judgment. Improperly selected diamonds, for example with excessively rounded points and edges, large point angles (exceeding 110°), heavy scale structure, cracks and inclusions close to the surface, adversely affect dressing tool performance. It has to be borne in mind, too, that the highest quality material need not necessarily be the most appropriate or economical to use in a specific application. The more perfectly developed a crystal is, the higher its price. For example, the price of rough diamonds suitable for single-point dressers depends in large measure on the number of cutting points and edges. Consequently, a number of diamond tool-makers supply dressing tools where the diamonds are in a range of quality grades. The best 'dressers' are those with perfectly developed diamond crystals having 4–6 points, with no fractures, pits or inclusions. When a cutting point wears, the stone can be dismounted and another point selected (Table 3.5).

Table 3.5 – Classification of diamonds for single-point dressers [198]

Kind of single-point dresser	Diamond characteristics
With rough diamond	0.2–4.5 ct crystals — octahedrons, rhombic dodecahedrons or spherical forms. No major inclusions or cracks visible under 3–5 × magnification Diamond quality: Quality 1 — 4–6 usable natural points, first class growth, no cracks, inclusions or pits Quality 2 — 3–5 usable natural points, good growth, no cracks or inclusions Quality 3 — 2–4 usable natural points, generally well-grown Quality 4 — 1–3 usable natural points, suitable growth
With polished diamond:	
— pyramid-shape dresser	Well developed octahedral or rhombic dodecahedral 0.25–0.75 ct crystals with flat or spherical faces, with distinct corner on quaternary axis of symmetry. Transparent crystals without impurities visible under 12–20 × magnification
— polished dresser (truncated or radial pyramid)	0.75–2.5 ct crystals without bubbles or impurities (which would make it imposible to cut or cleave the crystal to obtain elongated or flattened plates) visible under 12–20 × magnification. Edge chips may not exceed the thickness of the layer to be worked
— polished radial special profile dresser	Macle-type 0.25–1.5 ct crystals. No bubbles, impurities or cracks visible under 12–20 × magnification

3.1.3.2 Drillstones

Drillstones are a kind of natural diamond material used in oil and gas or mineral exploration drill bits, although the term does not cover all of the diamond types employed in such tools. A comprehensive listing of specialist terms is contained in the *Glossary of the Diamond Drilling Industry* compiled by A. E. Long [198]. Generally speaking, the term 'drillstones' is most frequently used to refer above all to crystals which are of oval shape and will therefore have a high mechanical strength. The sizes of these diamonds are roughly from 0.5 to 100 stones per carat (s.p.c.) and they are recovered on round-mesh sieves using a scale developed by the De Beers Company (the so-called diamond screens) or a scale originated by the Christensen Company (Fig. 3.4).

Drilling diamonds which naturally crystallize in a diamond (octahedron) type habit are highly sought after. The basic features characterizing these stones are blocky structure, obtuse angles between adjacent faces, and

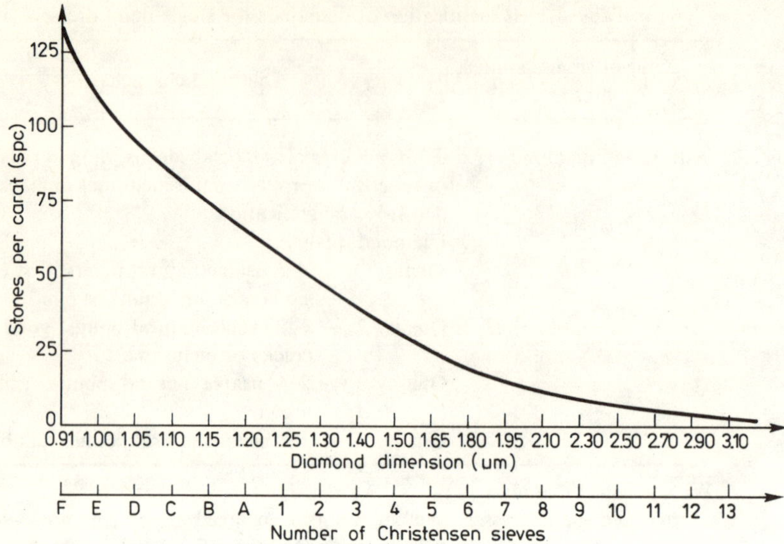

Fig. 3.4 — Approximate relationship between the size of drill diamonds expressed as the number of stones per carat and diamond stone dimension (diameter), and the round-mesh Christensen sieve number [44, 141].

ovalized edges and points. Ovalized octahedra and rhombic dodecahedra are found, with regularly crystallized cubes being somewhat less common.

The ovalized stones go by a number of names, such as Drill Rounds, A Grade, Premium, First Quality, Drill. Depending on how well they are crystallized, they are divided into a number of quality classes as listed in Table 3.6. The naturally ovalized isometric shapes are the most highly valued. At the other end of the scale is the so-called Casting, i.e. diamonds with sharp edges, not fully ovalized, often chips. This latter material is used in those parts of drill tools which are relatively little exposed to pressures and abrasion, but the main application is in so-called impregnated tools. An important role in diamond grading is played by the place of origin. Thus, for example, the so-called Congo stones, yellowish or greenish-yellow, mined in Zaire or Sierra Leone, show relatively low resistance to temperature and mechanical stresses. At the same time they are less expensive than the mechanically tougher colourless or brownish stones of the so-called West African (W. A.) type. For reasons of economy some use is also made of mixtures involving Congo type diamonds and higher quality crystals.

Stones recovered from worn tools, or imperfectly formed stones which have been subjected to rounding and polishing treatment, are known as 'processed diamonds'. A typical procedure consists of several

Table 3.6 – Diamond drillstones

	Name Symbol	Characteristics
Naturally shaped drilling diamonds	Select Quality, WA1, Drill, AAA	High quality, whole, sound and roundish, as mined
	Second Quality, WA2, AA	Medium quality, mostly whole (unbroken), fairly sound but may be somewhat off-shape and spotted. May include flat sided octahedra
	Third Quality	Lower quality, but sound and blocky enough for use in bits, has broken surfaces and sharp edges. An improved form is available that is made from diamonds that have been milled blocky and then lightly processed to a polish that follows the contour of the blocky diamond
	Casting	Material unsuitable for whole stone use is generally used in impregnated bits. Very flat plates and long pieces are excluded. High grade casting may also be made of Congo Boart that has been milled blocky
Processed drilling diamonds	Hardcore*, PQA1, Premium	Premium processed drilling material, produced by a series of shape improvement and surface treatment processes designed to enhance the drilling capability of selected natural diamonds
	Drill B*	A high quality processed drilling material of good shape and with a matt surface. The characteristics of Drill B are imparted by special processes carried out on selected stones
	ISD*, ISDS*	These two products are of more angular shape than Hardcore or Drill B. Selected diamonds are processed to give a product which is both strong and sharp

* Grades offered by De Beers.

stages of mechanical and chemical treatment. In the first stage the weaker stones are crushed and the remaining stones are ovalized and polished.

Processed diamonds are produced in various degrees of ovalization and surface polishing, the general rule being that drilling diamonds should have smooth surfaces. Reduced surface roughness lowers the friction, at the diamond/rock interface which, in turn, reduces the risk of diamond graphitization due to overheating. In widespread use are two ways of polishing diamond drillstones: mechanical and chemical. The chemical process involves dissolution (oxidization) of surface irregularities, while in the mechanical method they are removed by abrasion. In this way are obtained e.g. 'Hardcore' type diamonds, which are polished stones of approximately spherical shape [128].

Carbonado is a highly specific kind of drilling diamond. Polycrystalline carbonado 'stones' are porous and made up of a large number of fine diamonds. Properties vary but good quality material will have a high abrasion resistance, withstand dynamic mechanical stresses and also fairly wide temperature variations. Carbonado occurs in nature in the form of 'chunks' exhibiting different degrees of isometricity. Good quality carbonado is priced above the most expensive Drill Rounds and, because of that, it is used in special-purpose tools for drilling abrasion-resistant rocks. It is also used in conjunction with less expensive varieties of diamond, where the carbonado is set in those parts of the drill tool which are the most exposed to abrasive action [44, 169, 186].

3.1.4 Natural diamonds for special engineering applications

In engineering applications the internal structure, shape, size and surface state of diamonds are all considered, above all from the point of view of the mechanical properties of individual crystals — will they withstand the machining conditions? There is a large number of engineering products in the manufacture of which advantage is taken of the exceptional properties of diamond or where use is made of unique properties of the diamond crystal structure. However, for a particular diamond to do its job properly, it must be thoroughly examined, rejected or passed as 'fit' and prepared for the task.

The thermal conductivity of type IIa diamonds is several times better than that of any other known material (Chapter 1) and for that reason such diamonds are used for extracting heat in electronic devices — the so-called 'heat sinks'. Thermal conductivity depends on the purity of diamond, especially on nitrogen impurity: the presence of that element adversely affects that parameter. Diamonds to be used for heat sinks must be selected with particular care which means, above all, that each crystal must undergo rigourous physico-chemical instrumental analyses. Especially useful here is IR spectroscopy, for nitrogen impurity correlates with absorption in the 7–9.5 μm range. Having successfully passed the IR examination, a crystal is then tested for its thermal conduction parameters [72].

Typical heat sinks, manufactured on a commercial scale by e.g. D. Drukker and Zn (Amsterdam, Holland) are made as rectangular blocks, although circular, octagonal and 16-sided pieces are also produced (Fig. 3.5). Commonly used sizes are 0.25 mm × 0.40 mm × 0.40 mm and 0.50 mm × 0.75 mm × 0.75 mm. The two large parallel faces have an optical polish and flatness, as required for mounting the diode chip on the diamond and the diamond within the assembly. The side faces are generally less important and are left as cut with polishing. Heat sinks are produced by dicing plates and two methods of doing this are in use—mechanical sawing and laser 'cutting'. Laser cutting is more efficient and cheaper than sawing, but the side surfaces are somewhat rougher. Dimensional tolerances vary with the intended application. Thus, in the case of heat sinks designed for microwave diodes, tolerances of ± 0.05 mm on dimensions 0.5 mm and smaller are generally sufficient. For semiconductor laser devices and TWT helix supports, much closer tolerances are required [72].

Fig. 3.5 — Diamonds for heat sinks offered by D. Drukker of Amsterdam (courtesy of D. Drukker and Zn).

The high strength of diamond makes it possible to produce very thin diamond plates of comparatively large area. Thus diamond 'wafers' 5 mm in diameter and 15 μm thick have been produced for use as bolometer substrates. Thin diamond discs, wafers, etc. are made to go into various kinds of cutters and scribers used to make high precision cuts, grooves, or scratches. For example, D. Drukker and Zn [72] offers two groups of diamond products for the audio and video disc industries. The first group comprises rectangular logs, available in a wide range of sizes. Typical are logs 0.2 mm × 0.2 mm × 1.0 mm used as blanks for the production of styli for video disc players, or for making special audio styli. The optimum combination of diamond sawing, polishing and computer controlled laser cutting is used to produce logs to specified dimensional tolerances. Drukker also produces rough pointed logs, sharp pointed logs, metallized logs and ultra-small logs (0.1 mm × × 0.1 mm × 0.3 mm). The second group of products comprises a range of diamond styli for cutting audio and video disc “masters”. The cutting styli for audio and video disc “mastering” can be produced with sub-micron tolerances. The edge quality is checked with a scanning electron microscope (SEM) at magnifications of 20,000 × and over on a routine production basis. The logs and styli are oriented for optimum wear resistance by means of an advanced X-ray diffraction system which makes possible real time observation of the crystal orientation.

Another example of diamonds in the form of thin, elongated plates with profiled cutting edges at one end is the material used for the manufacture of surgical knives. The diamond plates are 0.15–0.3 mm thick, 0.1–3.15 mm wide and the length of the entire diamond, together with the cutting edge, is of the order of several milimetres. Such knives are made of gem quality diamond [72].

Owing to the wide range of electromagnetic radiation frequencies at which diamonds are transparent, some of the most homogeneous gem diamonds are used as windows in certain kinds of research apparatus. As in the cases described above, each crystal must be thoroughly examined and tested to see if it complies with the transmittance requirements. As a rule, diamond windows are made on a one-off basis with a particular application in mind. Their dimensions vary over a wide range. Since the parameter of interest is transmittance, diamonds of type IIa are used.

A situation similar to the one described above also holds true in the manufacture of diamond high pressure anvils. The shapes and sizes of selected diamonds are constrained by the design of the research equipment. Diamonds for this application are carefully selected and thoroughly tested for electromagnetic transmittance and mechanical strength. The final shape of the component is achieved by grinding, where account is taken of diamond anisotropy [80].

Ballas and carbonado are polycrystalline, naturally occurring diamonds. Both species, but especially carbonado, have found a number of applications in a variety of forms, from irregularly shaped chips (used e.g. in drill tools) to profiles (discs, wafers, etc.) cut from larger pieces.

3.1.5 Synthetic polycrystalline diamond tool blanks

Synthetic polycrystalline diamond is manufactured essentially in two grades:

— tool blanks with precisely specified dimensions,
— abrasive particles formed by comminution of larger pieces of blank (Section 2.3).

Synthetic polycrystalline diamond blanks, generally known as PCD (polycrystalline diamond) blanks, constitute a group of materials which are highly useful and very effective natural diamond substitutes in numerous practical industrial applications [49, 55, 56, 85, 92, 191]. Indeed, their properties being isotropic, in quite a few cases they are economically superior to natural diamond monocrystals.

Two basic kinds of PCD blank shapes are produced (Figs. 3.6 and 3.7):

— carbide-backed rounds, rectangles, squares etc. or segments thereof, made up of a thin layer of polycrystalline diamond bonded to a tungsten carbide substrate;
— polycrystalline diamond tool blanks with no tungsten carbide backing. Polycrystalline diamond tool blanks are marketed under various trade names which, at the same time, are registered trade marks. The leading manufacturers are De Beers (Ireland) and General Electric (USA). General Electric products include COMPAX,

Fig. 3.6 — Carbide-backed PCD tool blanks for use in the manufacture of turning and other machine tools (courtesy of De Beers).

Fig. 3.7 — PCD blanks for drawing die manufacture (courtesy of De Beers).

STRATAPAX, GEOSET and FORMSET, while De Beers PCD products include SYNDITE, SYNDRILL, SYNDIE, and SYNDAX-3. PCD materials are also made in the USSR [189], the most widely used known as BALLAS, CARBONADO, SLAVUTICH, SVSP, and ARS-3.

Indirectly, thermal stability decides the technology adopted for the manufacture of PCD tools as well as their use. From this point of view, PCD may be categorized into three classes [12]:

— material intended for use in tools made at relatively low temperatures, e.g resin bond grinding wheels. This type of material includes, inter alia, 'aggregate grain' sintered at normal pressures and at pressures approximating those of diamond synthesis;
— PCD blanks with 'medium' thermal stability, i.e. up to about 700° C

in a protective atmosphere. This group includes, above all, carbide-backed blanks for machining and for wire drawing, e.g. SYNDITE, COMPAX, SYNDIE;

— materials able to resist temperatures up to about 1420 K in a protective atmosphere. To this class might be assigned SYNDAX-3, ARS-3, FORMSET, GEOSET, and SVSP.

The first PCD blanks to be employed industrially were the COMPAX and STRATAPAX materials. Their properties are summarized in Table 3.7. COMPAX tool blanks are available for cutting tools, and COMPAX die blanks for wire drawing dies [49]. The material is distinguished by high impact strength, and the diamond part is especially resistant to abrasion. At the same time the carbide part is also resistant to spalling and cracking. COMPAX blanks can only be ground and lapped using diamond tools. The potential mechanical stresses which might arise due to the difference in thermal expansion of the two layers is to all intents and purposes eliminated by the use of an intermediate ductile metal layer between the diamond and the carbide and also by special packing of the diamond grains. The material should not be over-heated in air, the critical temperature being 950–1000 K. Above that temperature breakdown of the synthetic polycrystalline diamond may occur.

Table 3.7 – Properties of COMPAX, natural diamond and synthetic diamond [49]

Product	Relative abrasion resistance factor	Indentation hardness (kG/mm^2)
COMPAX	250	6500–8000
Natural diamond	96–245	8000–12000
Synthetic diamond	12	7000–8000

The De Beers counterparts of the two COMPAX materials are SYNDITE and SYNDIE. The properties of the De Beers materials are compared with those of natural diamond and ISO K10 carbide in Table 3.8.

A particular brand of PCD blank is usually manufactured in several grades using different diamond sizes (Fig. 3.8): in the commercially available blanks one can find particle sizes ranging from a fraction of a micrometre to about 120 μm [49, 55, 56, 85, 92, 191], coarser grained blanks being more prone to chipping. This can be seen from Fig. 3.9 where the attrition rates of different grades of SYNDITE are compared: as the diamond particle size increases, SYNDITE materials become

Table 3.8 – Properties of SYNDITE, natural diamond and K10 tungsten carbide [55]

	SYNDITE PCD	Natural diamond	Tungsten carbide (ISO K10)
Young's modulus (GPa)	841	964	630
Modulus of rigidity (GPa)	345	401	258
Poisson's ratio	0.22	0.20	0.22
Ultimate tensile strength (GPa)	1.29	2.6	1.0
Compressive strength (GPA)	7.61	8.68	4.51
Transverse rupture strength (GPa)	1.1	—	1.7
Hardness Knoop (GPa—at 20N load)	50	56–102	17.9
Thermal conductivity (W/m K)	560	600–2000	110

easier to fashion into tools because the voids between the sintered diamonds are relatively larger which makes it easier for the binder to penetrate into the material which, in turn, facilitates grinding of the blank. The pattern is quite similar in the case of COMPAX materials. On the other hand, the voids in the PCD adversely affect the finish of the work-piece surface. Also, the structure of the micropores influences the cutting rate.

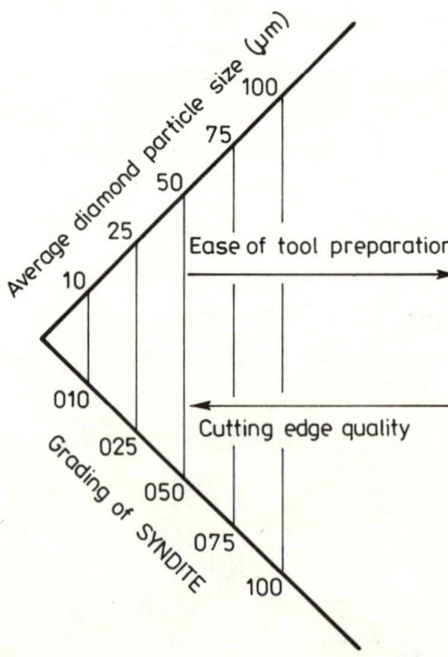

Fig. 3.8 — Useful characteristics and sizes of SYNDITE PCD blanks [1, 55].

SYNDAX-3, SYNDRILL, STRATAPAX, and GEOSET are specifically made for use in drill tools (Fig. 3.10). STRATAPAX is compared with natural diamond and C2 carbide in Table 3.9, while the properties of SYNDRILL are similar to those of SYNDITE. STRATAPAX and SYNDRILL are fully effective substitutes for natural diamonds, especially in drilling soft and medium-hard rocks.

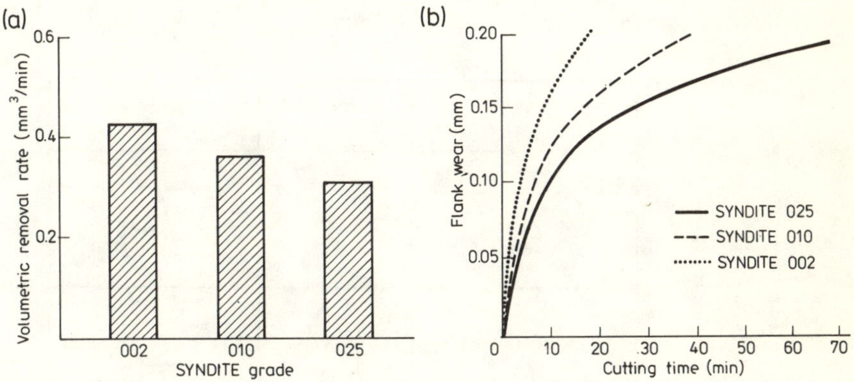

Fig. 3.9 — Comparison of cutting performance of SYNDITE PCD blanks differing in starting diamond grain sizes [1, 55].

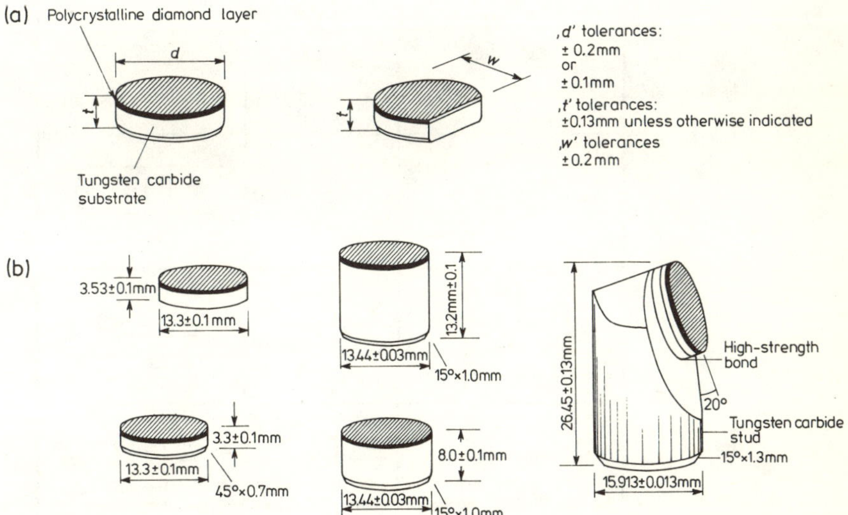

Fig. 3.10 — Typical sizes of STRATAPAX PCD blank shapes: (a) shape structure; (b) shape/sizes [191].

Table 3.9 – Properties of STRATAPAX PCD, natural diamond and tungsten carbide compared [191]

Product	Relative abrasion resistance factor	Indentation hardness (kG/mm^2)
STRATAPAX blank	200–300	5000–6300
Natural diamond	96–245	8000–12.000
C-2 cemented tungsten carbide	2	1,800–2,200

Product shape Shown actual size	Approximate carat wt./piece	Nominal dimensions (mm)
40° Triangle	0.12	4.0, 1.5
60° Triangle	0.28	4.0, 2.6
60° Triangle	0.90	6.2, 3.7
Cylinder	0.03	1.0, 3.0
Cylinder	0.50	3.0, 4.8
Rectangle	0.08	1.0, 1.0, 4.0
Rectangle	0.15	1.5, 1.5, 4.0

Fig. 3.11 — Typical shapes and sizes of FORMSET blanks [85].

The mechanical properties of PCD blanks for the manufacture of wire drawing dies, such as COMPAX and SYNDIE, are much the same as those of PCD used for machining. A very favourable factor which has significantly contributed to the widespread use of this kind of product is the ease with which the bore hole can be made by laser piercing and electro-erosion. Electro-erosion cannot be used on natural diamond as it is non-conductive. At the same time, the abrasion resistance of the PCD material is higher than that of (anisotropic) natural diamond, so that a greater quantity of wire can be drawn with a PCD die. Since it is possible to manufacture die blanks in large diameters, it is therefore also possible to draw wire of considerably larger diameter than with natural crystals. The blanks for wire drawing dies are made either with a carbide support ring or without one, in the form of monolithic cylinders, hexagonal blanks, etc.

Materials of the FORMSET and GEOSET type represent a fairly recent generation of PCD blanks, having been introduced onto the world market in 1982. FORMSET is used in the manufacture of wheel dressers, while GEOSET is for drill tools. These so-called thermally stable materials can be heated to 1450 K in an oxygen-free atmosphere. In air, FORMSET can be heated for 5—15 seconds up to no more than 1090 K. Prolonged exposure even to trace amounts of oxygen at this or higher temperature may cause breakdown of the PCD structure.

FORMSET is manufactured in prism, cube or cylinder form (Fig. 3.11). The cutting face geometry of FORMSET tools can be fashioned in the same way as with natural diamond. In contrast to monocrystalline diamond, however, FORMSET blanks wear gradually retaining the original shape of the cutting surface [85, 92].

Monolithic blanks of the FORMSET and GEOSET type also come with a thin metal coating, which provides additional protection against undesirable environmental influences at higher temperatures, leading for example to oxidization, binder penetration, etc.

Unlike conventional PCD materials, SYNDAX-3 has a ceramic binder thus avoiding thermal mis-match problems.

3.2 DIAMOND ABRASIVES

3.2.1 Natural diamond abrasives

Natural diamond abrasives, just like synthetic diamond abrasives, are available in sizes ranging roughly from 40 to 1600 μm (Fig. 3.12). The leading manufacturer and supplier of natural diamond abrasive is the De Beers company, with the Soviet Union taking a distant second place.

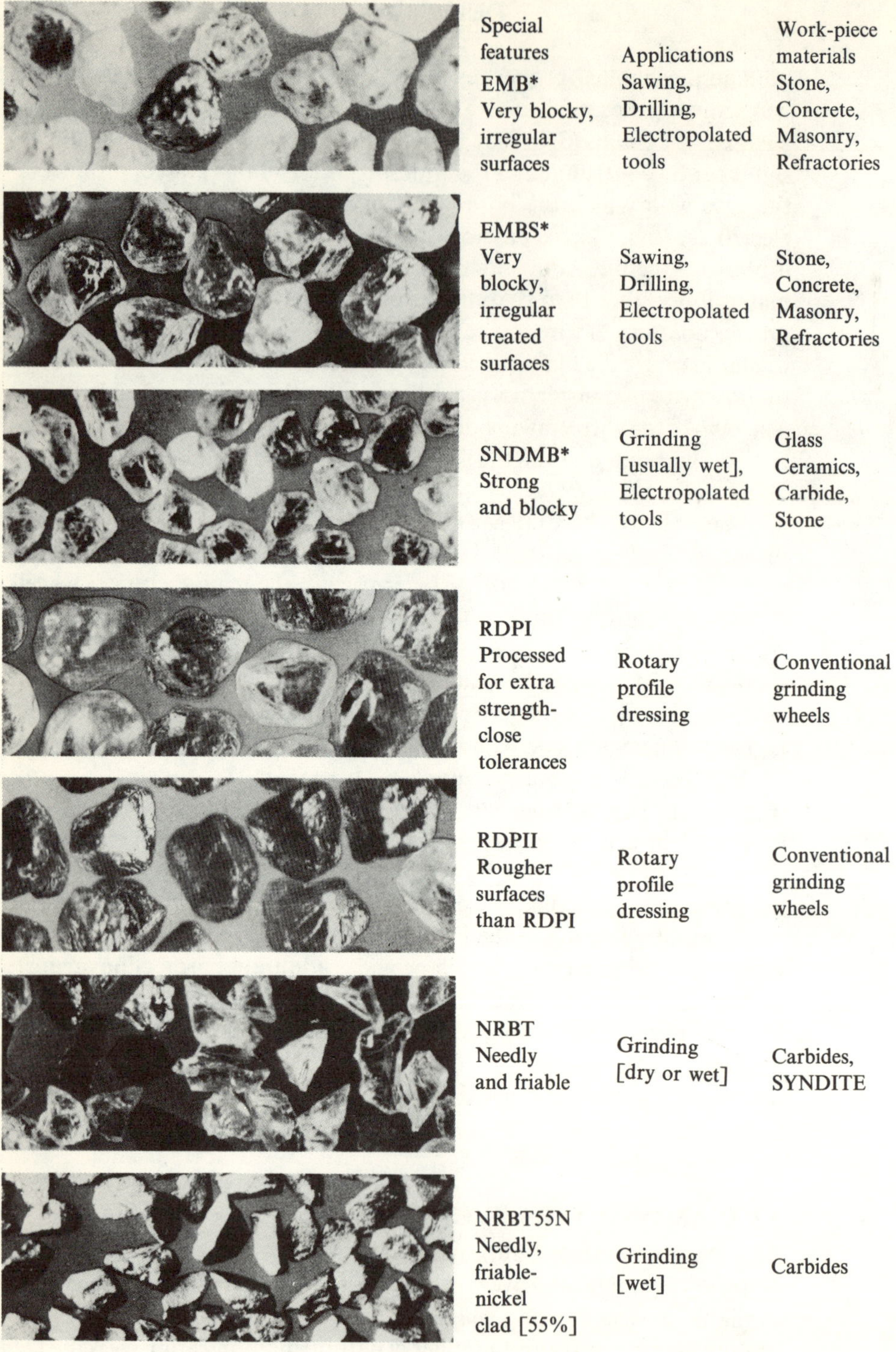

Special features	Applications	Work-piece materials
EMB* Very blocky, irregular surfaces	Sawing, Drilling, Electropolated tools	Stone, Concrete, Masonry, Refractories
EMBS* Very blocky, irregular treated surfaces	Sawing, Drilling, Electropolated tools	Stone, Concrete, Masonry, Refractories
SNDMB* Strong and blocky	Grinding [usually wet], Electropolated tools	Glass Ceramics, Carbide, Stone
RDPI Processed for extra strength-close tolerances	Rotary profile dressing	Conventional grinding wheels
RDPII Rougher surfaces than RDPI	Rotary profile dressing	Conventional grinding wheels
NRBT Needly and friable	Grinding [dry or wet]	Carbides, SYNDITE
NRBT55N Needly, friable-nickel clad [55%]	Grinding [wet]	Carbides

Fig. 3.12 — Different kinds of natural diamond abrasive offered by De Beers [54] (courtesy of De Beers).

Other producers of natural diamond abrasive base their products on the De Beers specifications.

Natural diamond abrasive is obtained by crushing and processing boart and defective stones of little commercial value, and to a limited extent by reclaiming the material left over from gem diamond processing [27, 54]. The diamond particles thus obtained are small, but they usually have no cracks or inclusions. Such particles are relatively strong mechanically and have a high thermal stability. Compared with synthetic diamond, they contain less metal inclusions and impurity atoms, so that the material is ideally suitable for the manufacture of sintered and electroplated metal bond tools (Table 3.10).

Table 3.10 — Comparison of natural and synthetic diamonds for electroplated tools [54]

Property	Natural diamond	Metal bond type synthetic diamond specially treated for electroplating
Iron impurity (parts per million)	10–50	1000–7000
Magnetic susceptibility ($\times 10^{-6}$ cgs units)	0–4	110–170

In the USSR [95], natural diamond abrasive is manufactured in five different grades, the divisions being based on the content of particles of isometric shape. Thus, diamond abrasive designated A1 should include no less than 10% of isometric particles, A2—20%, A3—30%, A5—50% and A8—80%. A5 and A8 diamond abrasives with the highest proportion of isometric shapes are recommended for use in metal bond tools used under arduous cutting conditions.

A somewhat different division into grades is used by De Beers [54]. NRBT for example is recommended for use in resin bond wheels for grinding heat sensitive sintered carbides. NRBT grit is distinguished by considerable brittleness and by largely non-isometric structure. Individual particles are frequently sword-like, flattened or elongated. They have sharp edges and points and they remain sharp even after a piece of the particle has chipped off. Particle sizes of NRBT diamond are in the 40–300 μm range.

Blocky and isometric diamonds carry the symbols SNDMB, EMB and EMBS. The former material is in finer sizes. These materials are strong mechanically and resistant to graphitization up to 1570 K. Low magnetic susceptibility makes SNDMB particularly suitable for use in electroplated tools designed for grinding glass, ceramics, carbides and minerals. Diamonds with similar characteristics to those of SNDMB but in sizes above 250 μm are designated EMB or EMBS. They are characterized by a blocky isometric structure and by an irregular surface, which makes for better keying of the particle in a metal matrix. The desired surface structure of EMBS particles is obtained by special mechanical-chemical treatment. The surface relief thus obtained is more abrasion resistant than an irregular type surface. EMBS diamonds are also relatively more resistant to mechanical loads and can work under tougher conditions, such as cutting and drilling of hard minerals, as well as being suitable for some types of dressers [54, 95]. The toughest of all, both mechanically and thermally, are natural isometric diamonds in sizes ranging from 50 to 400 s.p.c. which, after mechanical-chemical treatment, have been ovalized and their surfaces polished. Under very severe conditions, e.g. when used in dressers, rough surfaced and non-isometric particles break down relatively easily and fall out, while ovalized blocky particles are more resistant to mechanical pressures and thermal shocks. The premium De Beers natural diamond grades are designated RDPI, RDPII, and DEBDUST. These differ in the degree of particle surface smoothness and in the degree of ovalization; they are intended for use in dressers, saw blades and drill tools, and for protecting the surfaces of control-measuring instruments subject to wear.

3.2.2 Synthetic diamond abrasives

Synthetic diamond abrasives are designated by two sets of symbols which represent:

— manufacturer's designation,
— nominal mesh size.

The actual symbols and grade classifications adopted vary from one producer to the next (Flyleaf, Table 3.11). In practice manufacturers of synthetic diamonds follow the principle: "Tell us what you want to machine, and we will supply the right diamond". The divisions into diamond product types are intimately bound up with the kind of material to be machined (Table 3.11). The manufacturers suggest a particular diamond type as being optimal for a particular application in terms both of economy and surface finish. Not infrequently, the manufacturer will specify in great detail the application range of a synthetic diamond type; for example, De Beers recommend SDA100 or SDA100+ for sawing granite, SDA85 for sawing gabbro and syenite, and SDA100 for glass machining (Fig. 3.13). The grade classification of Soviet synthetic diamond abrasives is linked to compressive strength. The nominal mesh size is related to the dimensions or individual particles. The methods of checking the nominal mesh size (and size composition) of diamond abrasives are standarized (Fig. 3.14).

Table 3.11 — General Electric synthetic diamond abrasives [195]

Application	Work-piece material	Product family		Product type	Bond systems: Resin	Metal	Vitreous	Electro-plated
Sawing and drilling	Stone Concrete Glass Refractories Other non-metallics Aluminium oxide Silicon carbide	Man-made* diamond MBS*	Contains tough, blocky cubo-octahedral crystals with predominantly smooth faces. Transparent or translucent. Colour ranges from light yellow to medium yellow-green	MSD*		■		
				MBS-760*		■		
				MBS-750*		■		
				MBS-740*		■		
				MBS-70*		■		
				MBS-720*		■		
				MBS*		■		
				MBS-710*		■		■
Grinding and polishing	Glass Ceramics Hard metals Plastics Ferrite Fibreglass Dental enamels	Man-made diamond MBG*	MBG contains medium toughness, regular crystals. Colours range from yellow-green to light yellow, almost white. Crystal inclusion levels generally low, but some heavily included crystals evident	MBG-660*		■		■
				MBG-II*		■	■	■
				MBG-600*	■	■	■	■
				EBG*			■	■
	Cemented carbides Carbide/steel combinations Engineered ceramics	Man-made diamond RVG*	Friable, irregular particles, most frequently used with a metal coating, which totally covers all exterior surfaces. RVG-W, RVG-880 and CGS-II coated with a nickel-base alloy; RVG-D coated with copper	RVG*	■		■	
				RVG-W*	■			
				RVG-D*	■			
				RVG-880*	■			
				CSG-II*	■			

* Man-made is a Registered Trade Mark of General Electric Co., USA.

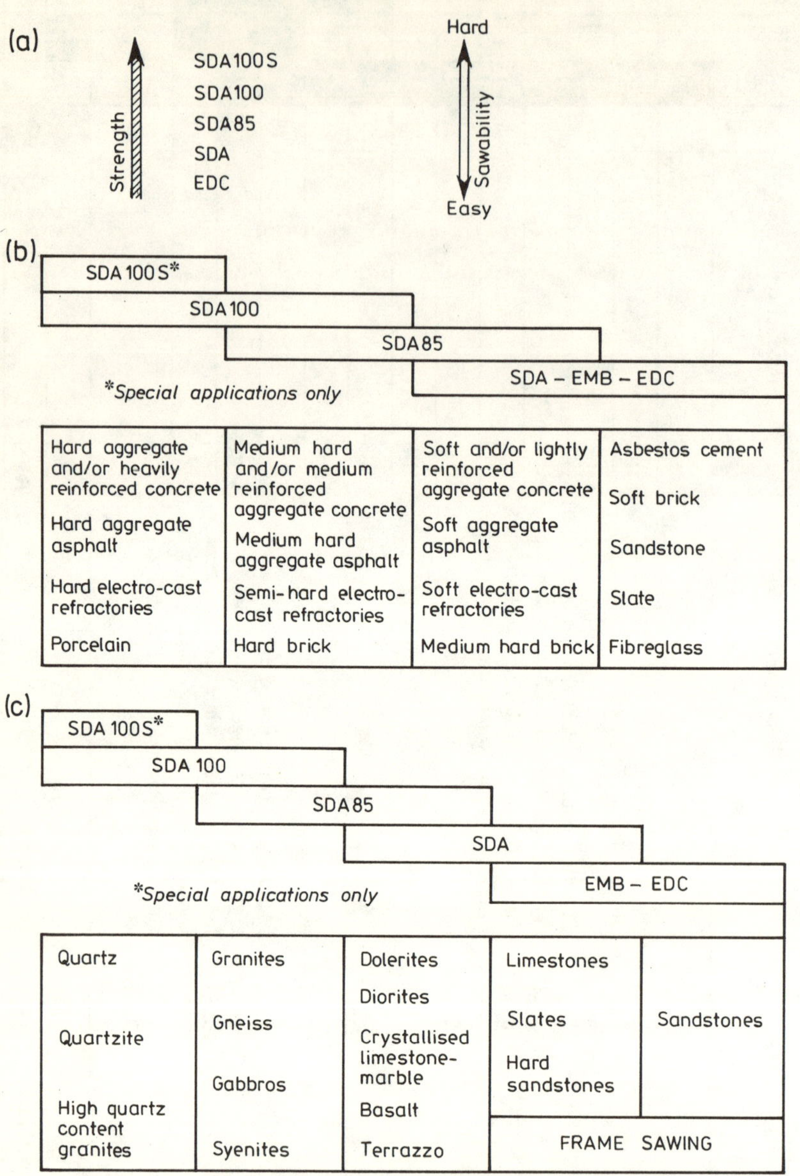

Fig. 3.13 — Properties of various kinds of De Beers diamond abrasives related to one another (a), and recommended applications in tools for working natural and synthetic construction materials (b and c) [54, 57].

The commercial grades of synthetic diamond abrasive are blends of crystals of different internal and morphological structure. Comprehensive assessment of the functional qualities of such crystal mixes requires physico-chemical tests which embrace:

— checking the size composition of the batch,
— analysis of the morphological structure and habit of randomly selected individual diamonds,
— determination of the parameters defining the mechanical strength of diamond particles, i.e. their resistance to static and dynamic pressures,
— tests on the thermal and electromagnetic properties of selected crystals.

There is a close relationship between the structure of a synthetic diamond particle and its physico-chemical and performance characteristics in tools [10, 28, 57, 95]. Based on microscopic observations, synthetic as well as natural diamond particles can be assigned to a number of groups characterized by crystal structure. In terms of internal structure and habit, the following particle types are distinguished:

— regularly formed monocrystals,
— monocrystals non-uniformly developed in various directions,
— polycrystals composed of coarse crystals,

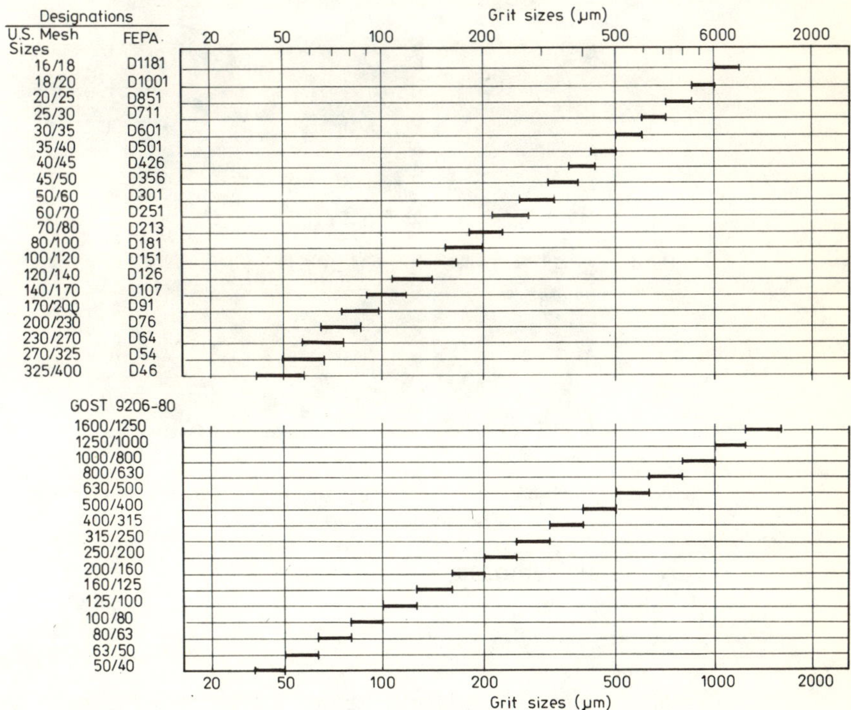

Fig. 3.14 — Approximate relation between FEPA, US ASTM and Soviet GOST grit size designations.

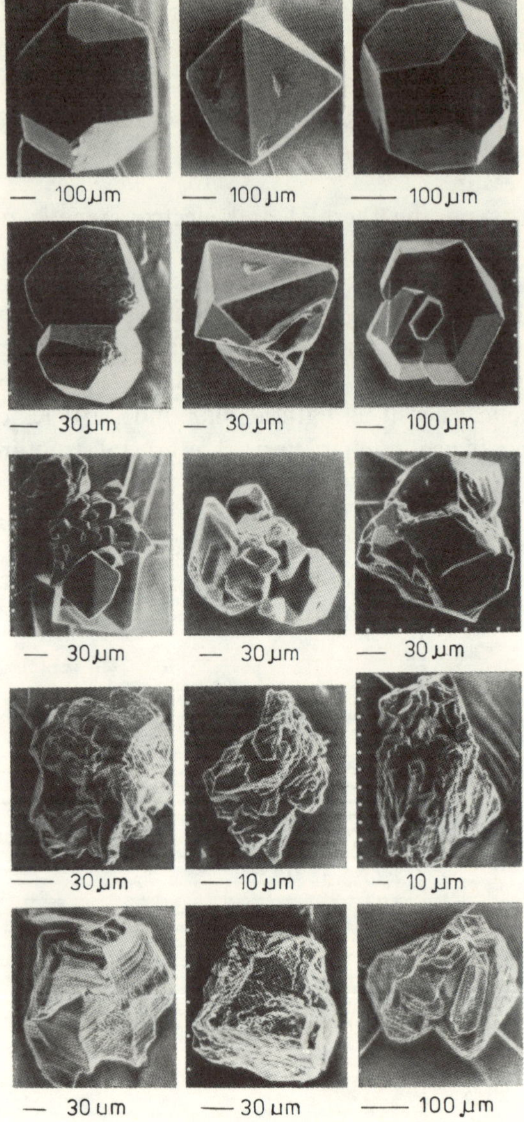

Fig. 3.15 — Characteristic of limiting forms of the structure and crystal habit of diamond abrasives [10].

— polycrystals composed of fine crystals,
— skeletal crystals,
— aggregate particles (Section 3.1.5).

Depending on the surface structure, the following particle types are distinguished:

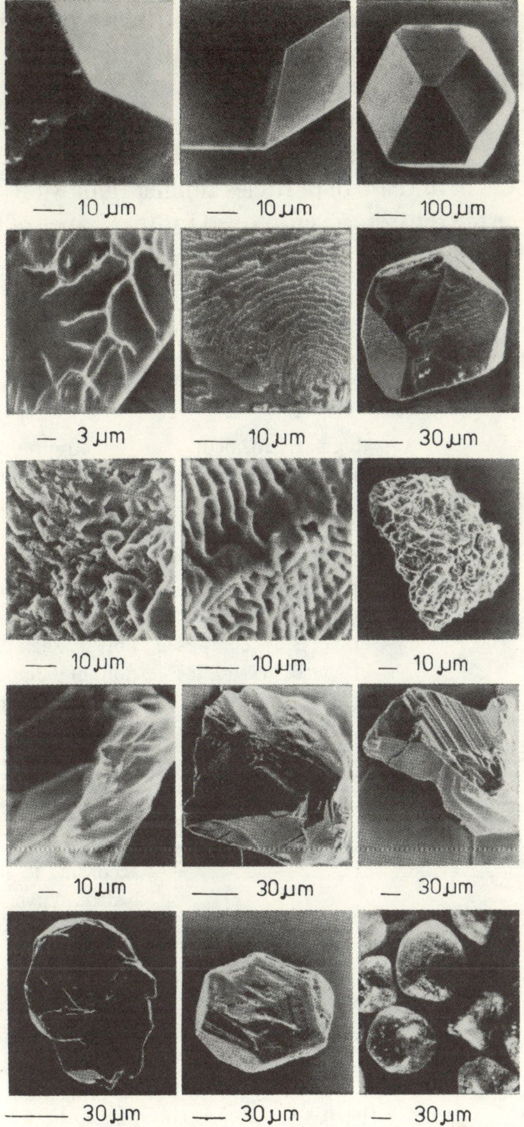

Fig. 3.16 — Characteristics of limiting forms of the surface structure of diamond abrasives [10].

— with plane, smooth faces,
— with flat, slightly irregular faces,
— with rough surfaces,
— crushed products,
— ovalized particles,
— clad particles (Section 3.2.3).

More details about the above particle types and examples of relative standards for each particle shape can be found in Figs. 3.15 and 3.16 and in Table 3.12.

The dimensional parameters of particles are evaluated by sieve analyses, and by microscopic comparisons of the extreme lengths of the contours in the case of particles smaller than ~60 μm. The description is made using the criteria employed for the shapes of typical abrasive grain [10]. The highest proportion (above 90%) of isometric particles is found

Table 3.12 — Diamond abrasive types [10]

	Crystal type	Characteristics
Structure and crystal habit of diamond abrasives	A — Regularly formed monocrystals	Isometric monocrystals such as octahedra, rhombic dodecahedra or combination of octahedra and cubic crystals Translucent, usually transparent
	B — Monocrystals grown unevenly in various directions	Monocrystals that are non-isometric, regular concretions, irregular concretions of some crystals > 100 μm
	C — Polycrystals composed of coarse crystallites	Polycrystals that are irregular concretions of diamond crystals smaller than 40 μm in diameter. Translucency dependent on impurities, colour and cracks
	D — Polycrystals composed of fine crystallites	Polycrystals that are irregular concretions of crystals smaller than 40 μm, but non-transparent
	E — Skeletal crystals	Crystals with layered structure, grown according to edges and vertices of octahedron. Layers of diamond divided with empty spaces, non-transparent or low speckled transparency
Surface structure	1 — Particles with flat, smooth faces	Smooth, flat walls of {111} or {100} planes. Sometimes on the surface small wavy cavities or etch pits visible at 300 × magnification
	2 — Particles with flat, rough faces	Flat walls with irregularities visible at 30 × magnification
	3 — Particles with a very rough surface	Very markedly developed surface of different character. Cracks, irregularities as result of concretions of small crystals or mechanical damage, pores or cavities may be visible
	4 — Crushed	Surface is the result of starting diamond crystal cracks
	5 — Ovalized	Particles with shaped or rounded edges and vertices

to be present in diamond abrasives recommended for metal bond and electroplated tools. This kind of particle could be approximated by a sphere whose surface is in contact with all tips of the particle. Diamonds of this kind are most frequently cubo-octahedra in varying degrees of habit development.

Elongated or nodular shapes will be found in diamond abrasives recommended for resin bond wheels and hones. Diamonds of AS2, RVG and CDA types, however, differ in shape from the special needle-like particle type CDA-L, formerly manufactured by De Beers (Section 3.2.3).

Extreme forms of diamond particle can be used for analysis of the composition of commercial types [156]. It follows from Table 3.13 that there is a relationship between the development of a single particle and

Table 3.13 — Comparison of the forms of diamond grain development and their application in tools [10]

Extreme forms of diamond grit development		Bond type				
Habit and inner structure	Surface	Metal		Electropolated	Ceramic	Resin
Diamond grain		M_1	M_2			
— Monocrystalline correctly developed	— with flat, smooth faces	×	O			
	— with flat, coarse faces	O	O	×		
	— breccia-like	×	×	×	O	O
	— ovalized	×	O	×		
— Not uniformly developed in all directions	— with flat, smooth faces	×	×	O		
	— with flat, coarse faces	O	×	×	O	O
	— breccia-like	×	×	×	O	O
	— ovalized	×	O	×		
— Coarse polycrystalline	— with flat, smooth faces	O	×		×	×
	— with flat, coarse faces		O		×	×
	— with very well-developed surface		O		×	×
	— breccia-like				×	×
	— ovalized					O
— Fine polycrystalline	— with very well-developed surface				×	×
	— breccia-like					O
— Skeletal	— with very well-developed surface					×
	— breccia-like					O

M_1 — Tools for rock, concrete, refractory materials, ceramics, corundum sawing and drilling, and for grinding wheel dressing.

M_2 — Grinding tools.

Grits with a given development form occur predominantly × or sporadically O.

the composition of samples of different types. At the same time, a comparison of the composition of samples and the application ranges recommended by the manufacturer indicates that the tool bond type and, by the same token, the severity of mechanical treatment likely to be incurred, determines the properties of the particles to be incorporated in the tool.

Microscopic examination does not allow unknown diamond samples to be compared in a sufficiently comprehensive way, and this also applies to their behaviour both during tool manufacture and during the work performed by a typical particle, when machining. The results of microscopic analysis of particles of unknown provenance should be supplemented by the results of additional tests. A very important rational parameter is mechanical strength. Strength tests are usually conducted using one of two methods:

- — measuring the compressive (quasi-static) strength of individual particles,
- — assessing the impact strength from the percentage of particles of the original size remaining after subjecting a batch of particles to a special test.

The compressive strength measurement of individual particles is the most important criterion of type classification of natural and synthetic abrasives of Soviet manufacture (Fig. 3.17). The technique of determining

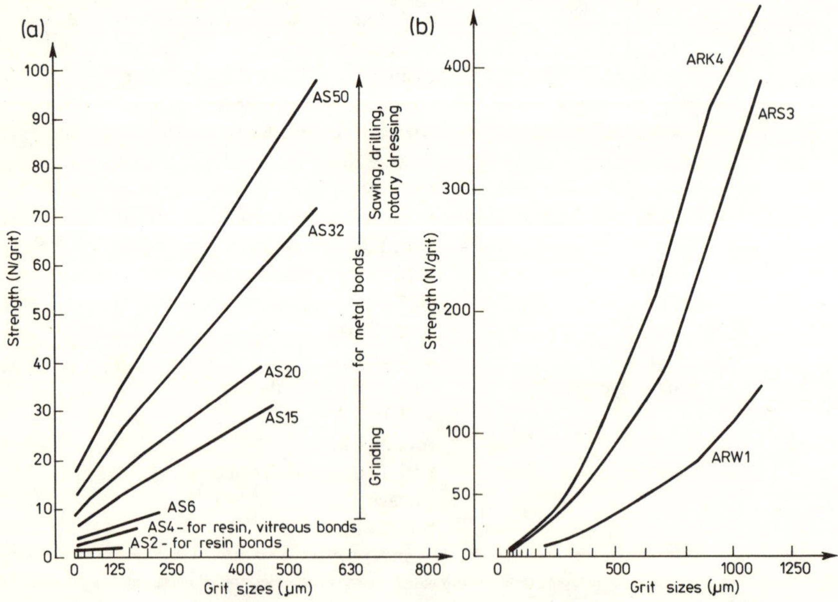

Fig. 3.17 — (a) Strength of Soviet synthetic diamonds [95]; (b) compressive strength of various Soviet PCD products.

the impact strength or friability of a diamond sample is a little more complicated. The most widely used device is the Friatester developed by De Beers [10, 80]. The results obtained by different measuring techniques are comparable. A given value of the "mechanical strength" of a particle predisposes it for working a given material. The most friable synthetic diamond particles exhibit highly developed and rough surfaces. They are used in resin bond grinding wheels and in honing sticks for machining sintered carbides, high alloy steels (honing only), glass and ceramic materials. The less friable abrasive types are used in sintered metal bond and electroplated tools, and also vitrified bond tools. As diamond is now recommended for increasingly more severe applications, an increasing proportion of strong, monocrystalline shapes can be observed in the samples. The strongest of all are crystals with a compact structure and smooth surfaces. They are used in sintered metal bond tools, such as saw blades, drill bits and wheel dressers.

The thermal properties of synthetic diamond particles may be evaluated in a number of ways. A typical test consists of heating a sample in an inert (e.g. argon) atmosphere at tool fabrication temperature, or at a temperature approximating the presumed working temperature of the particle tip. The purpose of the tests is to examine the changes taking place in the diamond crystal structure, specifically the degree of graphitization and cracking, and also to estimate any changes in the resistance of particles to mechanical loads. A diamond type is deemed suitable for a particular application if it does not undergo any changes during the test. Other thermal tests are used to investigate the behaviour of diamond particles in the presence of contiguous substances, e.g. the tool bond or air [10, 78, 80].

Synthetic diamond monocrystals exhibit the highest resistance to graphitization and oxidation (apart from natural diamond). These properties vary, however with particle size and surface condition. The thermal resistance of diamond decreases with inclusion and impurity content [10, 80]. Diamond types recommended for use in resin bond tools will burn in air more readily and undergo graphitization at lower temperatures than will blocky, monocrystalline diamond particles recommended for use in metal bond tools. For example, the diamond weight loss when heated for 30 min at 900 K is 1.6 wt % for ASW 60/40 GOST* particles, 6.1 wt % for ASO 60/40, 3.25 wt % for ASW 28/20 and 13.3 wt % for ASO 28/20 GOST particles [78].

Electromagnetic properties are investigated by analysing the be-

---

* The Soviet Standard for Diamond Abrasives specifies particle size fractions that are based on measurements in microns.

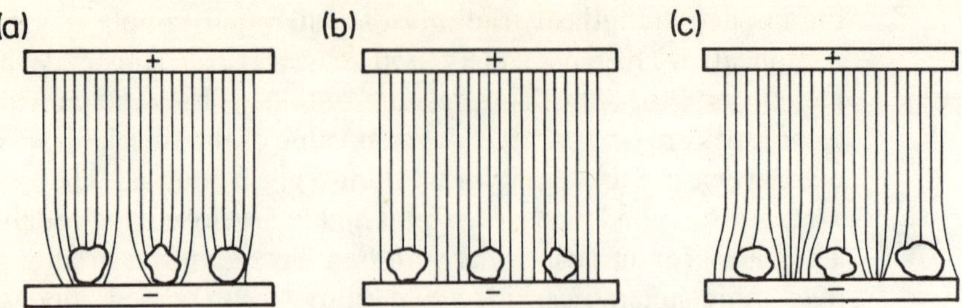

Fig. 3.18 — The influence of diamonds on the electric field when electrochemical bonding diamonds into tools : (a, c) diamonds changing the state of the electric field ; (b) the best diamonds for electroplated tools [54, 123].

haviour of diamond particles in electromagnetic fields, suitably chosen in terms of intensity. The particle behaviour depends on the kind and volume of inclusions, especially metallic, in the crystal. Special synthetic diamond abrasives for electroplated tools are manufactured only by De Beers (MDAE, MDASE) and General Electric (MBG-T, EP-100). Diamonds for electroplated tools should not change the state of the electric field between the electrodes in an electroplating bath (Fig. 3.18). Non-magnetic synthetic diamonds are obtained by special separation in an electromagnetic field and by chemical surface treatment.

3.2.3 Clad abrasives

In the course of abrasive machining using resin bond wheels, a large proportion of the diamond grits break down and pieces are pulled out of the bond without doing the job they are supposed to do. One technique for a more efficient utilization of diamond is to use clad particles, i.e. particles whose surfaces have been coated with a layer of metal or a non-metal several μm thick [54, 57, 83].

A number of advantages derive from coating diamond particles with a thin, compact layer, especially if it is metal : the (usually) polycrystalline particle is strengthened by the envelope, i.e. it is bound together more closely and coherently. In the course of abrasive machining, gradual breakdown of the particle takes place, which leads to self-sharpening of the tool. Moreover, the envelope protects the particle from oxidation over a given temperature range, improves bonding in the wheel. At the same time, the presence of metal at the diamond/bond interface slows down the rate of heat transfer from the diamond/work-piece interface into the bond, reducing the risk of thermal degradation of the bond surrounding and holding the diamond. Finally, owing to the natural ductile properties of metal, the particle is more flexibly secured in the bond and, consequently, it is more resistant to mechanical pressure arising during machining.

A variety of techniques for coating diamond abrasive particles are now in use, the most common being electroplating with metal, especially Ni, Cu and Ti (Fig. 3.19). Also used are glasses, organic substances and carbides. The cladding on diamond may consist of several layers of metal or several layers of metal and non-metal. Cladding can be achieved by different techniques, the most widely used being electrolytic and chemical, carbide formation, vacuum, cathode sputtering and vapour deposition, pressure spraying of a diamond-containing alloy and lastly, cladding during diamond synthesis [10, 83]. The cladding layers are deposited on single particles or on aggregates of several particles. One method of modifying the cladding process is to introduce another component, e.g. some other abrasive or diamond micron powder. The thickness of the cladding depends on the process parameters and on the diamond particle size.

According to [195], wheels with metal clad diamonds are used in more than 50% of cemented tungsten carbide grinding operations in the United States. In 1977, nickel or copper clad synthetic diamond or nickel clad natural diamond abrasives were widely used throughout the world in resin bond wheels. The nickel content varies between 30 and 56 wt % whilst for copper it is about 50% by weight. Clad natural and synthetic

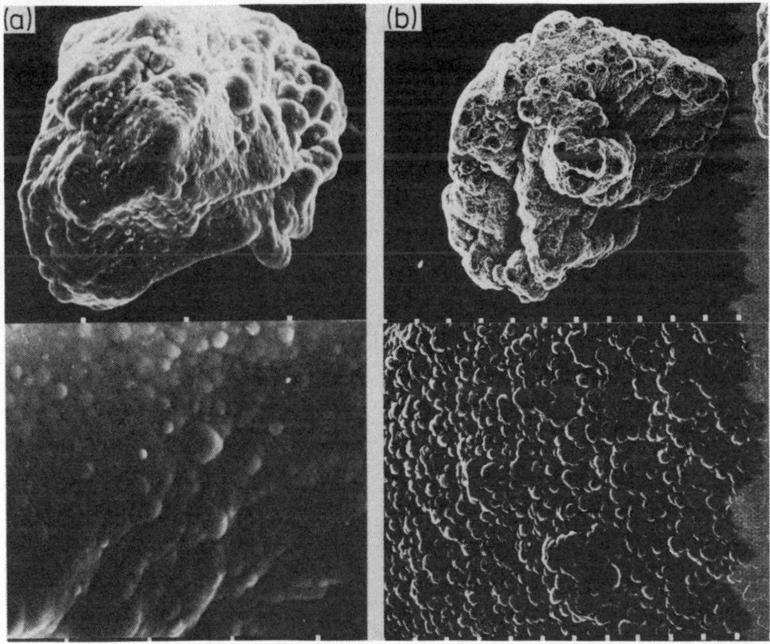

Fig. 3.19 — Metal-clad diamond particles: (a) with copper; (b) with nickel. Pictures at bottom show the structure of the metal cladding.

diamond abrasives for metal bond tools were put on the world market in 1978. The principal constituent of the cladding is titanium (EMB Ti clad and SDA Ti clad abrasives). The titanium coat is several μm thick. Metallization of the Soviet diamond abrasives ASO and ASR increases their compressive strength 1.5–2 times and the grinding efficiency of resin bond wheels containing such particles by 1.5–2.5 times [83]. The best results are reportedly obtained using metal alloys containing Sn and Ti.

A specific case of clad synthetic diamond particles is the former De Beers product designated CDA-L. The particles are of elongated needle-like shape, electrochemically coated with a nickel layer conferring magnetic properties [57]. CDA-L particles can be oriented in a magnetic field, thus making it possible to align particles optimally in a mould, i.e. at right angles to the grinding wheel face. As a result, advantage can be taken of the anisotropic properties of the diamond (Fig. 3.20). However, possibly on account of the cost of production, CDA-L abrasives are no longer available.

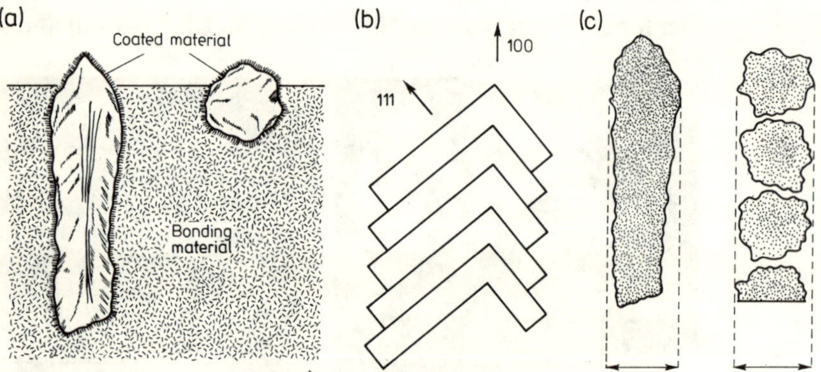

Fig. 3.20 — Synthetic, needle-shaped CDA-L type diamonds: (a) holding efficiency of needle-shaped particles in the bond compared with isometric particles; (b) structure of CDA-L; (c) comparison of particle utilization (needle-shaped vs. conventional synthetic diamond) [57].

In recent years, there has been an increasing interest in non-metal clad particles [10] (Fig. 3.21). Diamonds with a thin carbide coating are said to have a favourable effect on the life of metal bond tools. Diamonds coated with glass, devitrified material or metal oxides are reported to have an advantageous effect on thermal resistance when sintering vitrified bond tools.

Fig. 3.21 — Glass-coated synthetic diamond grit [10].

3.2.4 Aggregate abrasives

In addition to cladding of single particles, another widely adopted method of increasing the retention of diamond in resin bond tools is to use so-called aggregate grains (Fig. 3.22) [12]. Aggregate grains are made up of several, usually six or so, particles similar in habit, which are bonded together by a special alloy. The materials most frequently used for bonding are metals, metal carbides, glass and ceramics. Such metals as Ni, Ti, Cu, Co, Cr, Zr and their alloys, as well as other readily carbide-forming metals, are used.

An important advantage of aggregate particles is their irregular structure. Owing to the very rough surface, the particle is more firmly held in the bond from a purely mechanical point of view, compared with other kinds of diamond abrasive.

Aggregate diamond particles are not yet in widespread use but their importance grows steadily year by year. At the top of the list in this area in world trade is CDA-M (Carbide Diamond Abrasive—Multigrain) manufactured by De Beers (Fig. 3.23). The metal content is $55\% \pm 0.2$ g/cm^3. To date, this type of diamond abrasive has been manufactured in one size around 150 μm. It is recommended for the grinding of sintered carbides. Compared with conventional synthetic diamonds, the grinding efficiency obtained with CDA-M is 1.5–2 times as high [202].

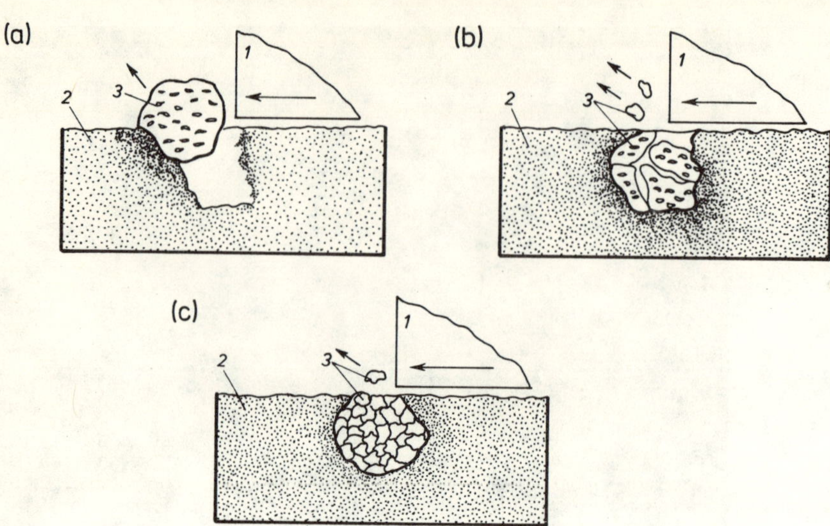

Fig. 3.22 — The mechanism of diamond grain chipping in the course of abrasive treatment: (a) monocrystalline grain; (b) polycrystalline grain; (c) aggregate grain; *1* — work-piece, *2* — grinding wheel, *3* — diamond grain [202].

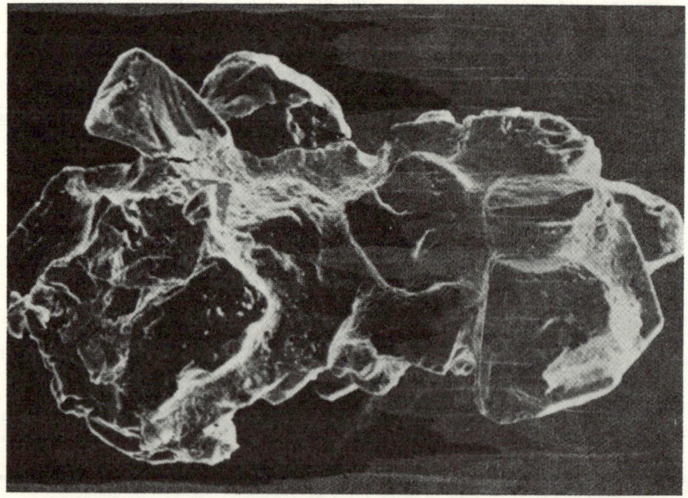

Fig. 3.23 — CDA-M diamond abrasive (courtesy of De Beers).

Aggregate grain of a different type is obtained by crushing polycrystalline diamond blanks (Section 3.1.5). Various Soviet products manufactured in this way are listed in Table 3.14, while Fig. 3.17b shows typical compressive strength values.

Table 3.14 — Aggregate abrasives made in the USSR [95]

Symbol	Characteristics	Applications
ARW 1	Obtained through comminution of synthetic 'BALLAS' type diamond	Honing of bores in cast iron, cutting glass and reinforced plastics
ARK 4	Obtained through comminution of 'CARBONADO' type diamond	Honing, working of stone and other construction materials
ARS 3	Obtained through comminution of 'SPIEKI' type diamond sinter	Manufacture of heavy-duty tools, e.g. bits for geological drilling, dressers, tools for working stone and other construction materials

3.3 DIAMOND MICRON POWDERS

Advances in technology, especially during the last fifteen years, have created a steadily increasing demand for diamond micron powders, i.e. natural or synthetic crystals less than 60 μm in size. This fine powder may be used as an abrasive, or as the starting material for the manufacture of polycrystalline diamond blanks [9].

The economic benefits accruing from the use of micron diamond are strictly determined by the morphology and internal structure of individual particles — their shape, mechanical strength, surface characteristics and linear dimensions. They also depend on the uniformity of individual particles in a batch and on the way in which they are used. The quality of micron diamond powder depends on the raw material and the manufacturing conditions, the products offered by different manufacturers showing considerable variety. The structures of natural diamond and of diamond particles obtained in static synthesis less than 20 μm in size are much the same. These are essentially blocky shaped crystals. Coarser particles obtained by static synthesis have a compact monocrystalline-clastic structure, although particles with flat, smooth edges can also be seen in the samples. The percentage of particles of this kind increases with the diamond size. Micron particles obtained by dynamic (or shock wave) synthesis exhibit a different structure, resembling an agglomeration of many very fine crystallites (Section 2.2.4) distinguished by a considerably greater number of cutting edges. The use of this kind of diamond micron powder can greatly reduce lapping and polishing time.

The utility of a diamond micro powder is reflected in its abrasion and polishing properties, i.e. the weight of material removed per unit time, and the surface finish of the workpiece. Figure 3.24 shows the values for surface finish and abrasive efficiency as laid down by the Soviet standard GOST 9206-80 for Soviet-made diamond micron powders. The leading diamond micron powder manufacturers, with top output, produce a material with the so-called normal abrasive-polishing properties. In addition to standard products, they also manufacture micron powders featuring enhanced functional properties, obtained by chemical treatment of the diamond surfaces and by more precise shape and size segregation. Diamond micron powder with improved abrasive-polishing properties is manufactured mostly for specific machining applications.

The admissible differences in the size and structure of individual particles of diamond micron powder are very small when compared with the requirements for other kinds of diamond material. Separation of particles so much similar to one another requires not only special apparatus and techniques, such as the Amplex Diamond Auto-Grader (USA), but also a high standard of purity of the starting material. The presence of a single 'harmful oversize' particle in a batch may produce a

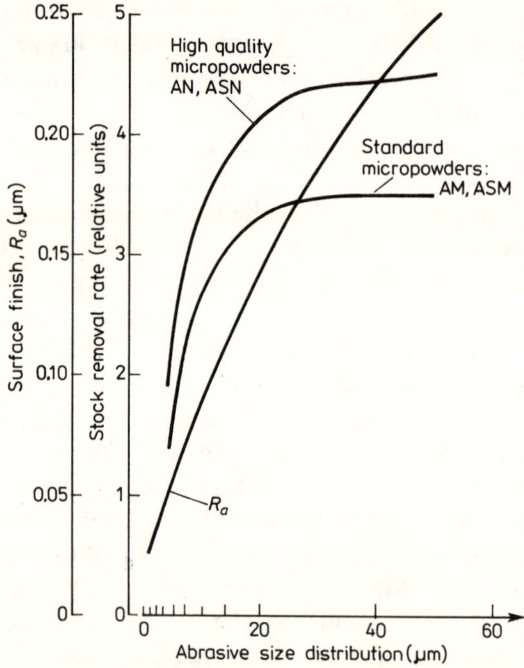

Fig. 3.24 — Surface finish and stock removal rate using Soviet synthetic diamonds [117].

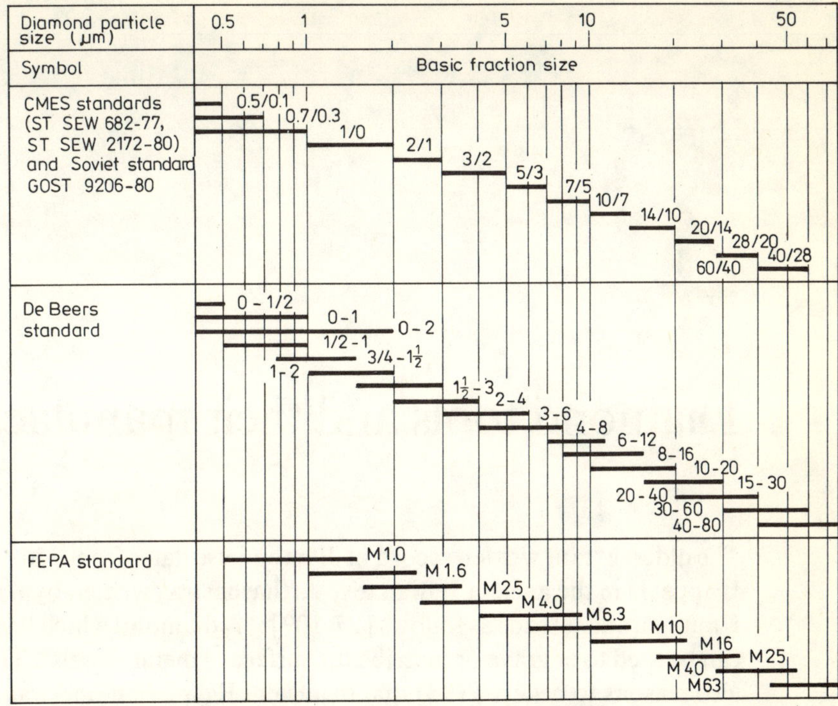

Fig. 3.25 — Range of micron diamond sizes.

scratch on the surface of the workpiece which will be extremely expensive to remove. Various national standards exist governing the sizing and descriptions applied to micron diamond powders. However, marked differences exist between the international FEPA standard and the standards of diamond manufacturers (Fig. 3.25). Micron powders in sizes below 1 μm are called sub-micron powders.

4

Diamond tools and their manufacture

The oldest extant written record of diamond use dates from 350–300 BC. It appears in the ancient Indian text 'Arthaśhastra', written by Brahman Kautilya, and it reads as follows [27, 109]: "A diamond which has points can be used to engrave or inscribe the surface of metal vessels". There are also reasons to believe [198] that in ancient Egypt diamond was used to make tools. Still, it is only since the middle of the 19th century AD that one can seriously talk of the industrial production of diamond tools, and of designs which have led to the current applications. The first diamond-set drilling tools were made by Hermann and Leschot, then came circular saw blades (Fromholt), diamond-tipped lathe tools (Ramsden, Dumaine), and diamond wire drawing dies (Milan, Balloffet, Brockendon). The first diamond grinding wheels were made in the early 19th century (Pritchard, 1824) but they remained a rarity until the early 20th century (Czapski, 1906; Cauthier, 1927). Resin bond diamond wheels appeared in the 1930's (Wickman, Stanford, Voegli-Jaggi), i.e. at a time when rapid progress was taking place in plastics technology, and sintered tungsten carbide tool and die products were becoming increasingly more widely used.

The extreme hardness of diamond and its unparalleled abrasion resistance continue to make it eminently suitable for all manner of industrial applications. In accordance with Mohs' principle diamond, the hardest known material, may be used to cut (or sever), abrade (remove in small pieces), or plastically deform any other material which does not chemically react with it excessively fast. However, for any kind of mechanical working using diamond or, for that matter, any other material, to become possible, the single diamond particles must act on the work-piece material with a certain energy, and the action must be appropriately oriented in space. As dictated by economic considerations, the relatively small dimensions of diamond crystals, abrasive particles or micron powders generally makes it necessary for them to be mounted in holders, sintered into abrasive segments, suspended in fluid carriers, etc. Owing to the vast diversity of workpieces and work materials, an almost infinitesimal variety of diamond tools and compounds has been developed.

The design and dimensions of a given tool depend directly on the nature of the machining operation, the characteristics of the work-piece material, and the surface finish requirements. In terms of the quantity of diamond involved, tools may be divided into single- and multi-point varieties. In multi-point tools the diamonds may, but need not be, oriented relative to one another and relative to the work-piece surface. The diamonds may be specially shaped or they may be used in their natural state. Manufacturing methods include:

— tools in which the diamonds are mechanically mounted,
— tools with brazed in diamonds,
— tools made by powder metallurgy methods,
— tools with diamonds bonded by electrodeposition,
— vitrified bond tools.

The methods employed in the manufacture of products other than tools, e.g. lapping compounds are broadly similar.

Depending on the character of diamond action on the work-piece material, three main types of tool can be distinguished:
— tools for abrasive machining and polishing,
— tools for machining,
— tools for plastic and cold working.

Diamond tools can also be sub-divided into types related to application, such as grinding wheels, cutting tools, wire drawing dies, drill bits, wheel dressing tools, etc.

4.1 SHAPING DIAMONDS FOR SINGLE-POINT TOOLS

Diamonds for single-point tools are used as supplied (Chapter 3), or the rough crystals are appropriately shaped to designated tip radii and angles between the faces, or they are ovalized, hollowed, cut into plates, etc. Shaping is required in the case of natural diamonds or PCD tool blanks prior to mounting in such tools as wheel dressers, hardness testing indentors, gauge points, wire drawing dies, etc. Also, diamonds to be used in products other than industrial tools, e.g. jewellery. For most engineering and non-engineering applications, the methods of diamond working are similar.

The hardness anisotropy of diamond makes it necessary to precede any mechanical operation with a thorough examination of the crystal morphology and internal structure, as a basis for deciding the optimal course of action, including the polishing directions. Based on years of experience, especially that accumulated by generations of manufacturing jewellers, a number of simple and effective instruments and techniques for diamond polishing have been developed [27, 130, 175]. The first stage is removal of redundant or defective parts of the crystal, which is effected by cleaving or sawing.

A diamond crystal to be cleaved is first cemented onto a special rod called a dop, after which the place where cleaving is to be initiated is selected and marked (scratched) with another diamond crystal or a laser beam. Cleaving is effected with a steel wedge which is struck with a steel rod. Cleaving should be done along the {111} planes and the cleaver must possess expert knowledge of diamond crystal structure, have excellent sight and a steady hand — any error during this highly critical operation may cause shattering of the crystal. Though very slow, diamond sawing involves fewer risks and for that reason it is generally the preferred preliminary shaping operation. The recommended sawing directions are shown in Fig. 4.1. As can be seen, they correspond to the {100} or {110} planes.

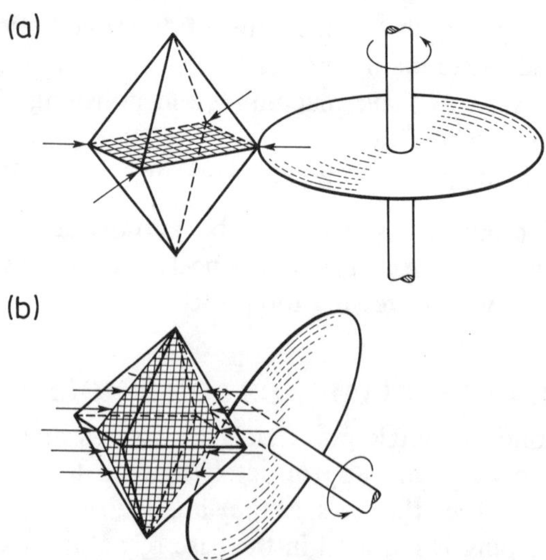

Fig. 4.1 — Recommended diamond cutting directions: (a) octahedron cutting in the {100} plane; (b) octahedron cutting in the {110} plane [27, 139].

Diamond crystals are cut with thin copper alloy discs or blades whose periphery is charged with micron diamond powder. Depending on the size of crystal to be cut, saw blades of different thickness are employed. For example, 0.03–0.04 ct diamonds should be sawn with blades 0.04 mm thick, while 1 ct stones should be sawn with blades 0.065 mm thick, etc. [63]. It is good practice to use a succession of increasingly thinner saw blades. A stand for diamond sawing is depicted in Fig. 4.2. As a rule, blade spindles are driven by transmission belt at around 11,500 r.p.m. The diamond to be divided is mounted in a special holder (called a tang) that is designed in such a way that the crystal rests on the edge of the saw blade and the pressure exerted by the diamond on the saw blade derives from gravity and is adjustable.

An operation known as bruting consists in imparting a circular shape to the base of the stone. The turning stone is set in a holder and mounted in the chuck of a lathe. The actual shaping is done with another diamond which abrades and roughly rounds the base or girdle of the stone, rotating together with the holder in the lathe chuck.

The shaping of natural diamonds, both industrial and gem quality, requires a good 'feel' for the crystallography. Appropriate orientation of a crystal is often aided by high precision optical instruments, or better still X-rays [130, 175]. No such orientation problems exist when grinding* PCD materials, the mechanical properties of which are isotropic. For this very reason, however, PCD materials are difficult to grind. At the same time it is quite easy to automate PCD blank polishing. The blanks are mounted in holders from which they are removed when lapping and polishing has been completed. The blanks are brazed or clamped into a toolholder in a separate technological process. In an alternative procedure, a toolstone is mounted in a tool shank and only then is its edge given the required shape. In the case of natural diamond, grinding and polishing are effected using cast iron wheels (called scaifes) rotating at 1500–3000 r.p.m. The scaifes are charged with micron diamond particles or they are saturated with diamond compound. Any run-out or vibrations when the scaife is rotating will have an adverse effect on work-piece accuracy. Grinding progress is monitored at high magnification. The scaife surface is usually divided into three zones:

— an internal ring used for locating — by trial and error — the most

---

* Carbide-backed PCD blanks are made by the producers in the form of circular pieces up to ~70 mm diameter. These discs are then cut up into any practical combination of size or shape using wire EDM. Any grinding process referred to by the authors would be a subsequent stage — Translation Editor.

Fig. 4.2 — (a) Diamond sawing machine; (b) complete diamond sawing bench (courtesy of Rubin and Son, Antwerp).

suitable point on the crystal face from which to start grinding (the so-called 'trial' zone),
— a middle ring for grinding to the required dimensions (the so-called 'grinding' zone),
— an outer ring for polishing the crystal face, which is where the fine diamond micron powder or compound is usually introduced (the so-called 'polishing' zone).

The three zones may also come in other sequences. The transition from the grinding to the polishing ring is made when the former no longer produces any improvement in surface finish. Owing to the hardness anisotropy of diamond, it is imperative to constantly maintain proper orientation in terms of 'soft' and hard directions. An experienced diamond polisher* recognizes a good grinding direction by the sound. As a rule, a properly oriented diamond crystal produces a soft sound on contact with the scaife. Another practical indicator of orientation may be the surface quality obtained. The harder an improperly oriented stone is pressed, the less accurate are the grinding results. Flat surfaces, as well as surfaces of revolution (conical, cylindrical and spherical) may be shaped by grinding and polishing using specially designed machine tools, where the diamond work-piece may be rotated or which provide for stepless adjustment of the grinding angles and directions.

It is recommended that PCD tool blanks should be machined using specially designed machine tools such as the one depicted in Fig. 4.3. Suitable machinery is manufactured, inter alia, by Ewag AG (Switzerland), Wendt GmbH (West Germany), Coborn Machine Co (UK) and Supreme Diamond Tools Ltd (UK). The grinding results are significantly affected by the grinding wheel characteristics. For example, the Winter company of Germany manufactures POLYBLOC wheels where a special binder composition is used [218]. As a rule, grinding is carried out in 2–3 stages. For example, in the case of SYNDITE PCD blanks the Ademant company of Germany recommends a two-stage process (rough and finish grinding) [1]. Wheels should be of the 6A2 type, about 150 mm in diameter with a face width of about 20 mm, rotating at 20–30 m/s. The best results are obtained with vitrified bond wheels containing De Beers CDA diamond, but resin bond wheels with NRBT or SND Micron diamond as well as metal (iron-bronze) bond wheels with MDA diamond may also be suitable. The optimum diamond concentration is 75–100. The recommended grit sizes are about 400 US

* The term 'diamond polishing' is traditionally used to include both grinding and polishing — Translation Editor.

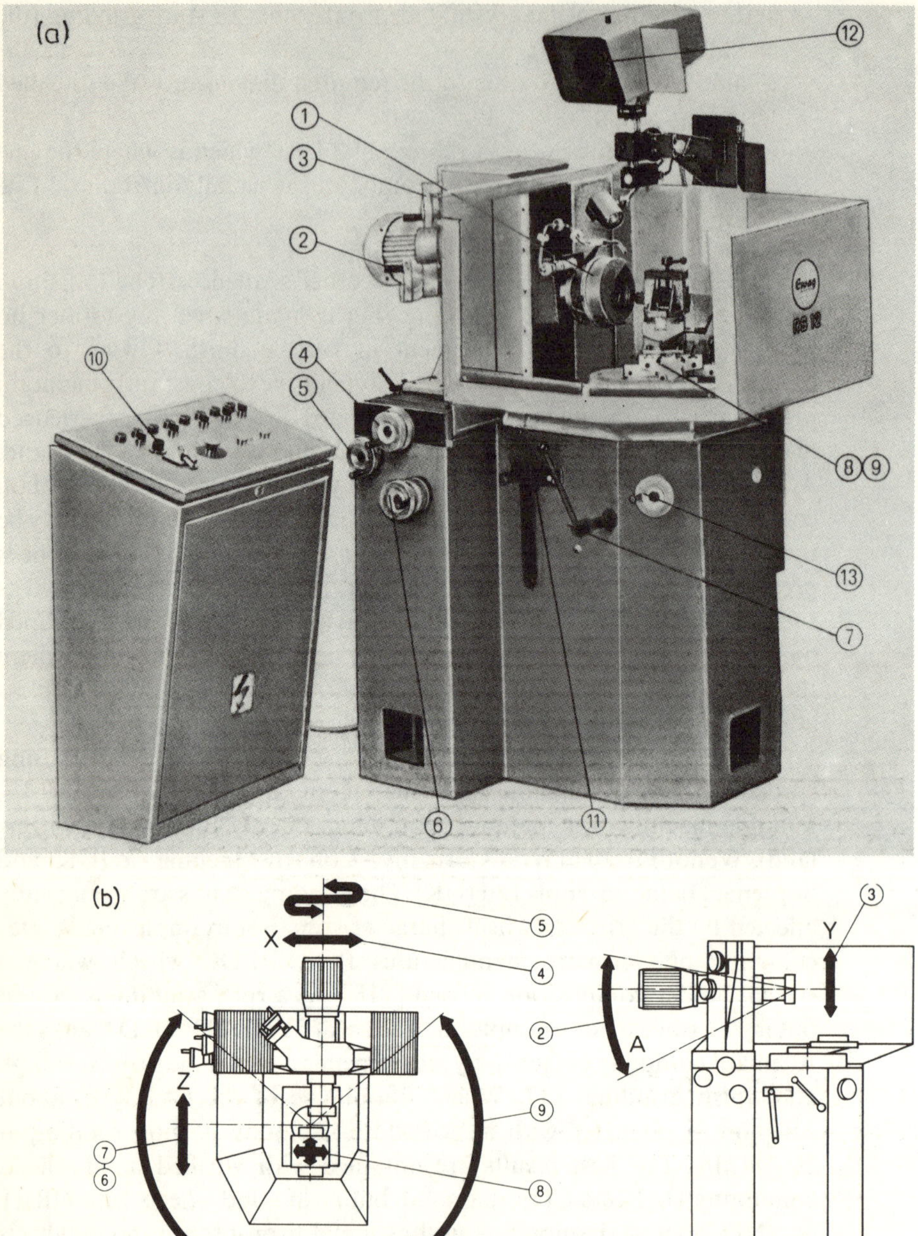

(a)

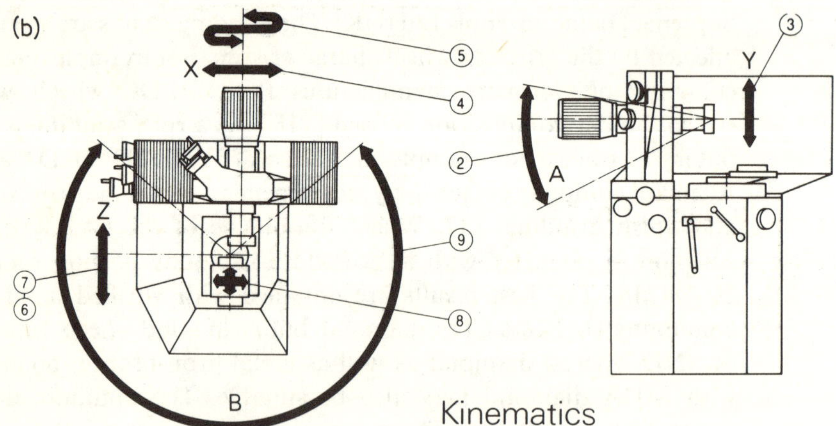

Fig. 4.3 — Typical high-precision grinder for PCD tools with optical control of the machining process (exemplified by RS12 machine manufactured by EWAG) (a), and the range of movements possible in the process (b) (courtesy of EWAG Company, Switzerland).

mesh for rough grinding and 8–16 μm for finish grinding. When grinding COMPAX PCD blanks, the best results are achieved using resin bond wheels containing RVG-W diamond, the recommended grit sizes being 170/200 US mesh for rough grinding and 325/400 US mesh for finish grinding. Whenever a very fine surface finish is required and, by the same token, a sharper and more uniform cutting edge, PCD tool blanks are lapped and polished on planetary lapping machines. Wheel wear is quite extensive. When rough grinding SYNDITE PCD blanks, the G-ratio (volume of PCD removed relative to the volume of wheel wear) is in the 0.01–0.02 range. The PCD removal rate attainable is then up to 0.5 mm^3/min. Such rapid wheel wear makes it imperative that the wheels be examined at frequent intervals, so that their cutting faces can be trued and dressed whenever the need arises.

In addition to those indicated above, there are other methods of shaping diamond crystals for single point tools. One such process, described in the chapter on wire drawing dies, is the piercing of holes in PCD blanks. Yet another method is ovalization, which yields nearly spherical crystals. The process is conducted for a large number of crystals at the same time, and it consists in the diamonds abrading one another in air-jet mills or in planetary lapping machines, cascade-rotational lapping machines, etc.

4.2 MANUFACTURE OF DIAMOND TOOLS

The range of uses for diamonds when made into tools is partly determined by the physico-chemical properties of the tool bond and above all its resistance to heat, chemicals and mechanical loads, as well as its wear resistance. It has also to be kept in mind that each of the various bond types makes a good tool only with diamonds of strictly defined structure (see Chapter 3).

4.2.1 Tools with brazed-in diamonds

Brazing as a method of mounting diamonds in tools is employed above all for single-crystal and PCD tools. Brazing causes relatively rapid heating of the diamond or PCD to a high temperature, which carries the risk of the crystal or tool blank cracking or of surface oxidation. Heat is applied using a blow torch or electrical resistance or induction coils. The latter method is generally best. To reduce the risk of the diamond crystal getting damaged in the process, use is made of brazes with melting points of less than 1070 K. The most popular brazes for diamonds are those based on silver; they hold up well under the conditions of tool use and, at the same time, they have wettability and bond well with both the diamond and the steel tool shank. Brazing also involves the use of fluxes, e.g. a mixture of borax and boric acid or boron oxide and fluorides.

More sensitive to elevated temperatures — as compared with natural diamond — are PCD blanks, especially if they have a tungsten carbide substrate, such as those used in cutting tools. High temperature may cause the blank to crack due to the difference in thermal expansion of the two layers. Such PCD materials can be kept at temperatures in excess of 870 K for no more than a few seconds (Fig. 4.4). The brazes recommended

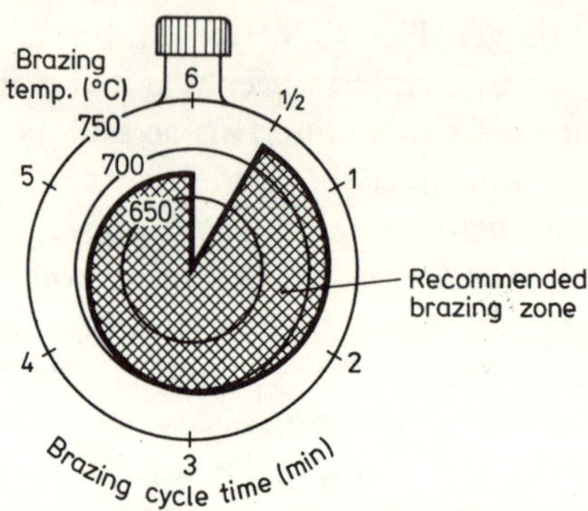

Fig. 4.4 — Recommended times for brazing SYNDITE PCD blanks in tuxing and other cutting tool shanks [55].

for use with SYNDITE PCD tools are listed in Table 4.1. In accordance with General Electric recommendations [49] the best brazing filler metal for COMPAX PCD tools is Handy and Harmon Easy-Flo 45. This silver brazing alloy melts at 880 K and flows at 895 K, well below the temperature at which thermal damage to the PCD occurs. Easy-Flo 45 is available in sheet form 0.13 mm thick. Before brazing, wafers of Easy-Flo 45 that conform to the size and shape of the blank to be brazed are cut from the sheet. If large numbers of identical wafers are required, an investment in a suitable punch and die set will pay off. The rectangular strip of brazing alloy that fits into the back of the pocket should always be cut so that one edge projects from the end of the tool. By observing

Table 4.1 — Brazes recommended for use with SYNDITE PCD [55]

Product	Manufacturer	Temperature (°C)	
		Solidus	Liquidus
Easy-Flo No. 3	Johnson Matthey Metals	620	630
5009	Degussa AG	645	690
Easy-Flo No. 45	Handy and Harman	632	688

this edge the operator can tell when the alloy has reached brazing temperature. Handy and Harmon-Flux — a paste — is recommended when brazing COMPAX PCD tools. An advantage of this flux is that it becomes transparent at 870 K, the temperature at which it is fully active. For brazing, heat can be applied either by a torch or an induction coil. The advantage of an induction coil is that the temperature can be controlled more precisely. Torch brazing is usually done with an oxy-acetylene torch, although other fuel gases also give satisfactory results. The procedure for induction brazing is similar to that for torch brazing. The induction coil is positioned behind the cutting edge of the tool in such a way that the heat flow to the pocket is at a maximum. The heat source is removed as soon as the brazing filler metal starts to flow. An advantage of induction brazing, in addition to good temperature control, is that this procedure requires minimal operator skill and training. Graphite fixtures are often used for induction brazing of COMPAX PCD tools. They expedite the brazing process and are easy to fabricate. Finally, a word of caution — heat must always be applied to the tool shank or holder, never directly to the PCD blank. This will minimize the possibility of thermal damage to the blank.

4.2.2 Tools manufactured by powder metallurgy methods

There are three basic technologies employed in the manufacture of diamond tools by powder metallurgy processes:

— infiltration,
— hot pressing,
— cold press and sinter.

These techniques are employed above all in the manufacture of diamond abrasive tools, although they can also be effective in bonding single crystals or PCD blanks. The above processes can also be used to prepare 'semi-products' for the manufacture of tools, in the form of blanks where the diamond is held in sintered metal powder. These blanks (or biscuits) are then brazed to the tool shank.

In terms of diamond layer structure, tools with sintered metal bonds fall into two categories — surface-set and impregnated. Surface-set diamond tools are fabricated in moulds where the diamond crystals are cemented at strictly predetermined locations, following which the mould is filled with bond material and the whole is sintered. Especially useful for this purpose are the infiltration and hot press techniques. Diamond impregnated tools, on the other hand, are made by sintering a fine diamond and bond mixture, with the diamond particles dispersed throughout the entire layer. Thus, in surface-set tools there is a single

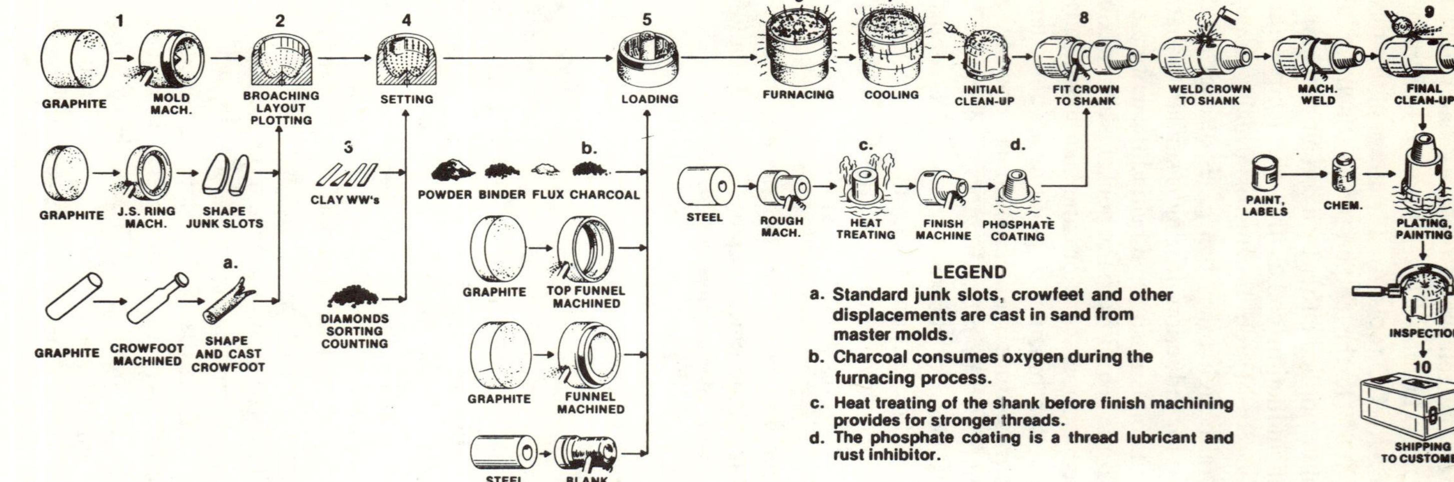

Fig. 4.5 — Activities involved in the manufacture of a diamond tool (exemplified by a geological drill) by the infiltration method [44]. The numbers refer to the following operations: *1* — a concave carbon mould is machined from a graphite cylinder to give the general shape (or profile) of the bit, *2* — spherical-shaped indentations for the diamonds are drilled on the inside surface of the mould and the vertical hole-gauge surface and drilled or broached into the mould surface, *3* — fluid course displacements made of sand, carbon or clay are placed in the mould, their purpose being to provide suitable channels in the finished bit for proper fluid distribution between the diamonds, *4* — the inside of the mould is then coated with a rubber-based glue and the correct size and quality diamonds are individually set, using vaccum needles, into their respective machine indentations, *5* — a steel blank is centred in the finished mould to be bonded metallurgically to the crown, formed from tungsten carbide powder which is poured into place between the carbon mould and the steel blank, *6* — the mould assembly is placed in an electric furnace where the binder metal melts and infiltrates by capillary flow into the hot powder mass, *7* — after a controlled cooling cycle, the matrix materials solidify permitting the crown to be removed from the carbon mould, *8* — the crown casting is machined and threaded and an API tool joint connection (the shank), made from a heat-treated alloy steel, is permanently affixed to it, *9* — final steps include API ring gauging, identification stamping and final inspection.

layer of larger diamonds while in impregnated tools, the diamond particles are much smaller and can be incorporated throughout a layer theoretically of any depth or height. All three of the above-mentioned processes can be employed in the manufacture of diamond impregnated tools.

In the infiltration process, a porous skeleton is first made and diamond particles and high melting point metal powder are then introduced, to be bonded together when a liquid bond flows into the pores. The successive stages of tool fabrication by the infiltration process are depicted in Fig. 4.5, the example being a drill bit. Materials typically employed to obtain the porous skeleton are carbide powders (e.g. of tungsten, molybdenum, titanium, chromium) and metal powders (e.g. tungsten, chromium, molybdenum) as well as easy-to-sinter metal powders (e.g. cobalt, iron, nickel, bronze, copper) [33, 42, 125, 157]. In order to obtain sufficiently dense packing, the powder in the mould may be vibrated. The bond alloy, which has a comparatively low melting point, is contained in a small reservoir with an outlet adjacent to the matrix powder surface so that upon melting, the bond alloy is drawn by capillary action into and throughout the compacted matrix powder. Pressure may or may not be employed subsequent to the infiltration process, depending on the type of tool and the properties desired. The infiltrating agents are usually copper alloys, e.g. Cu–Ni–Zn, Cu–Ni, Cu–Sn, Cu–Mn–Ni, Cu–Ag. Heat treatment takes place in box furnaces in a protective atmosphere, or induction heating is used to a temperature of 1250–1500 K; the maximum sintering temperature should be at least 100 K higher than the melting point of the alloy used for infiltration. The latter will flow more easily and fill the intergrain pores more thoroughly if fluxes are employed. In the case of tool bonds based on tungsten carbide and tungsten, the higher the melting point of the infiltrating alloy (e.g. Cu-Zn-Ni), the higher the resistance of the final product to abrasion [223]. The bond quality is also influenced by the size, shape and surface structure of the particles making up the porous skeleton. Thus, particles with regular shapes and smooth surfaces make better contact with the infiltrating alloy and, consequently, produce stronger bonds [83]. Good examples of such powders include Descarb tungsten or Macrocrystalline WC type tungsten carbide supplied by the Macro Division of Kennametal, Inc. (Canada). The infiltration method is ideal for the production of suface-set tools because it minimizes the forces that would dislocate the diamonds from their set positions in moulds. There is essentially no dimensional change after the powders are placed in the mould.

The hot pressing process consists in heating a diamond impregnated metal bond mixture accompanied by pressing. In the third variant, carefully filled moulds are first cold pressed and then heated in special ovens until they reach the furnaces sintering temperature. In many cases, protective atmospheres are provided. Sintering is followed by judicious heat treatment, which is of the utmost significance for the structure and consequently also the properties of a metal bond grinding wheel. The alloying agents are included with the powder mix which, after compaction in the mould, is either pressed after sintering or heated simultaneously to the sintering temperature. The above two techniques produce different physical properties with the same matrix powder. Hot pressing is a single-stage process and is therefore quite fast. Good quality control is generally assured. Figure 4.6 shows an apparatus for fabricating

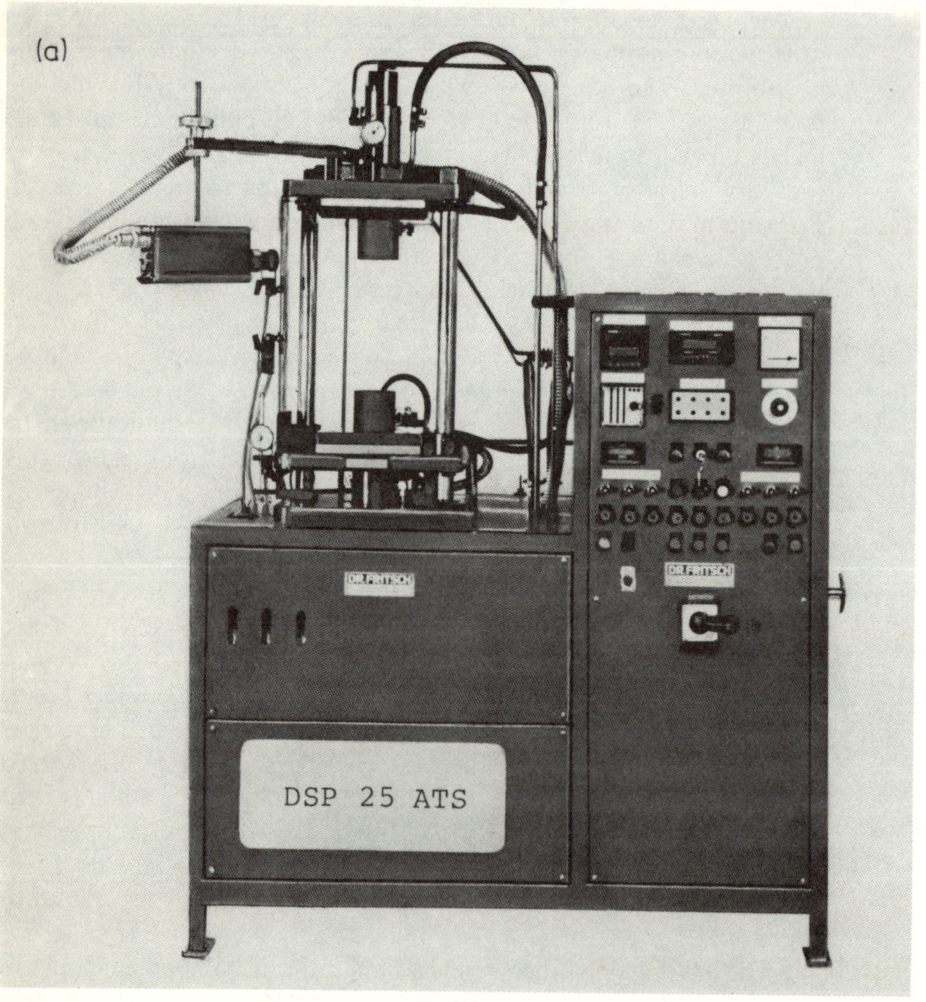

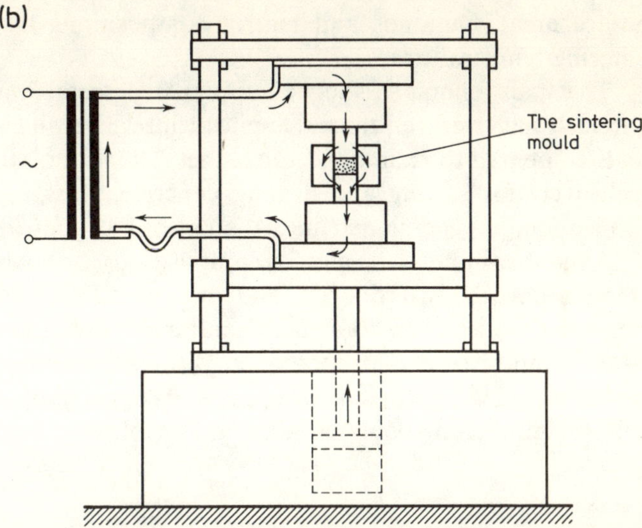

Fig. 4.6 — (a) Typical press for making diamond tool segments by hot pressing of metal powders, exemplified by the 'Sintering Press' manufactured by Dr Fritsch, Germany (courtesy of Dr Fritsch KG, Fellbach, FRG); (b) schematic diagram of the 'Sintering Press' apparatus and direction of current flow through the graphite mould at the time of tool sintering [89].

diamond tool elements by the hot press process. The time and temperature sintering parameters are automatically controlled. Heating may be by electrical resistance or induction. In the former method, heating occurs when the electrical circuit is shorted as the two halves of the mould make contact. The temperature is usually measured pyrometrically. Presently marketed systems, for example from Dr. Fritsch (Germany), Boliden Finemet (Sweden) or Workdiamond (Italy), are both simple to operate and highly efficient. The apparatus may be operated in conjunction with other equipment, e.g. diamond powder + metal powder blender and mixer, a system for brazing diamond segments to the tool body, a device for final preparation of the tool, etc.

The cold press and sinter process, like the sintering and hot press methods, uses self-bonding powders but involves a different sequence of steps. Self-bonding powders are usually supplied with an organic bond material which becomes fugaceous at fabrication temperatures but serves to give temporary or "green" strength to the compacts [89, 126]. The powders are formed into tool shapes by pressing in moulds and are then removed from the moulds to be placed, free standing, in an atmosphere-controlled oven for sintering. The cold press and sinter process allows the use of comparatively rapid mechanical pressing. Hydrogen or other reducing gases ensure an oxygen-poor environment, which is important for minimizing oxidization especially of small (high surface area) diamonds and matrix compositions that require high sintering temperatures.

The main characteristic of impregnated metal bond tools is an extremely high wear resistance. Despite their relatively high manufacturing costs compared to resin bond tools, metal bond tools are used almost exclusively for sawing and drilling concrete, glass, stone, etc. Their application, in general, demands the use of a coolant.

A number of bond (or matrix) compositions for floating die and hot pressing have been developed, and there is no universal, all-purpose blend. In practice, every tool manufacturer employs his own, thoroughly tested compositions and sintering conditions. Metal bond constituents in widespread use are based on copper and copper alloys, iron, cobalt, as well as high-melting point carbides and metal powders (Table 4.2). The

Table 4.2 — The influence of various components of metal bonds on bond properties in diamond tools manufactured by powder metallurgy

Main bond component			Examples of tools		
Melting point	Approximate abrasion resistance	Metal	Slow wearing bond		Non-abrading bond — the working element is the protruding diamond point
High ↑	High ↑	Tungsten carbides Silicon carbide Tungsten Molybdenum Cobalt Nickel Iron Copper Brass Bronzes Silver and silver solders Aluminium Zinc Tin	Dressers Drilling bits Saws	Grinding tools	Tools with brazed diamonds
↓ Low	↓ Low				

maximum sintering temperatures for bronze bonds are in the 1000–1250 K range, while for those with high carbide and tungsten contents — in the 1200–1550 K range. The maximum compaction pressure depends on the mould strength. Weighed amounts of diamond and matrix material are cold pressed in steel moulds using pressures from 0.3 to 7 T/cm^2 [220]. Hot pressing up to about 1150 K may be carried out in steel moulds. At higher temperatures, graphite, ceramic or special heat-resistant alloys should be used, the most popular being graphite moulds, typical designs of which are available from specialized companies, e.g. Gravatom Industries Ltd (UK).

A simple but popular bond material is a copper-tin mixture, often modified with other elements. According to Wendt Report No. 324, the most advantageous copper-tin weight ratios for diamond abrasive tools are from 3:2 to 4:1. Bronze bonds are in wide use in Europe, Japan and in the Soviet Union (Table 4.3). For example, in the USSR a bond material with copper and tin powders (weight ratio 4:1) is designated M1, and the optimum sintering temperature—in terms of maximum tool life—is 970–1090 K, with a compaction pressure of 2.5–3 T/cm^2 [157]. After

Table 4.3 — Properties of metal bonds used for Soviet diamond wheels [188]

Bond designation	Mechanical strength (MPa)			Brinell hardness
	Bending	Compressive	Shear	
M1	610	815	87	89
MS1	303	530	71	91
MS6	369	610	76	91
MK	334	590	83	110
M04	140	481	15	82
M15	260	540	63	113
M013	159	461	36	97
MW1	150	471	34	88
TM2	117	491	18	84
M5-5	116	481	39	91

sintering, the bond is an alloy of tin in α-Cu and in $Cu_{31}Sn_8$ (δ phase). Copper and tin are also the basic constituents of Soviet bonds designated MS1, MS6, M15 and MK. In the M15 bond there is an addition of silver which forms an advantageous eutectic mixture with the δ phase [10]. In the MW1 bond, copper and tin are present in combination with low melting point metals such as Mg, Al and Zn [157]. Also common are bond materials without any tin, e.g. Cu–Zn–Al in which the main crystal phase is $CuAl_2$ and, less usually, a solid solution of Zn–Al with a trace of copper [90]. In wide use also are compositions where, in addition to a metal powder other substances are employed as so-called fillers. These may be fine ground glass, silicon carbide, alumina, etc. The presence of carbide and high melting point metal fillers (e.g. Ti, Cr, W, Zr, Mo) in tool matrices increases the abrasion resistance [223]. To reduce friction, graphite is added to the bond composition. On the other hand, bonds with phosphorous and nickel additives are recommended for lapping and polishing. Bond compositions may also include inorganic salts, e.g. chlorides, fluorides, sulphides and bromides. Those containing fluorides are particularly suitable for grinding high lead content (i.e. crystal) glass [151].

Bonds based on cobalt and metal carbide powders, and also high melting point metals, are sintered at temperatures above 1320 K. The use of iron in diamond tool manufacture is limited by its tendency to react with carbon (diamond) at elevated temperatures. Nevertheless, thc presence of ferrous metals usually improves the resistance of a tool bond to abrasion [29, 197]. Bond materials as thus discussed are recommended for tools for rock drilling, and for abrasive wheel dressers and drill bits. Bonds with the highest abrasion resistance are based on Co–WC (or W), having melting points above 1270 K. The diamond particles to be sintered with metal powders may be clad or unclad; in the former case, titanium metal is the most common (Chapter 3).

4.2.3 Diamond electroplated tools

Another member of the metal bond tool family is the electroplated bond. The physico-chemical processes in E. P. tool production can be described in terms of Faraday's Law. An example of apparatus used in the manufacture of such tools is shown in Fig. 4.7. A finished tool consists of a metal hub with diamond particles adhering to the cutting face. The

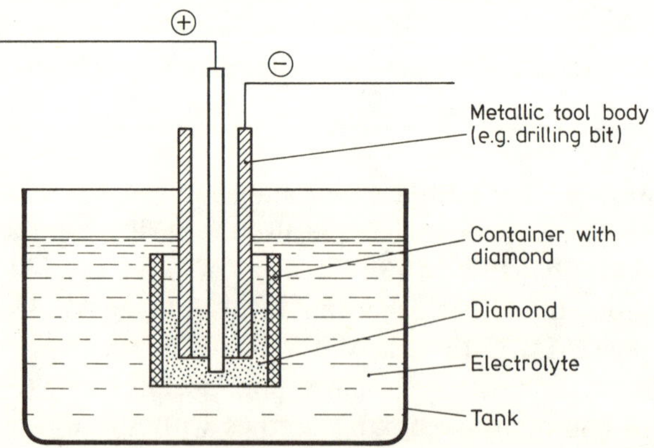

Fig. 4.7 — Schematic representation of the equipment for diamond tool manufacture by electroplating.

toolmaking process consists in pouring diamond powder onto the working face of the tool in an electrolyte bath. The rest of the tool is 'masked off' with e.g. laquer. The metal depositing onto the tool under the influence of electric current 'coats' the crystals, and in this way bonds them to the tool hub or blank. The metal most frequently employed is nickel. Nickel can be deposited in Watts type baths or from sulphide-chloride solutions. The basic compositions of the two baths are quite similar, the differences in the results being related to the chemicals added to the solutions. The types and amounts of additives are usually protected by patents. It is for the most part the additives that determine the quality of the electroplating. In the process of diamond bonding, use is often made of an extra layer (hard nickel or chromium) as well as a coat (copper). The resistance of electroplating to abrasion can be improved by adding other abrasives. However, the efficiency of E. P. tools is significantly influenced by the quality of the diamond employed. The particles should be similar in size and they should have a blocky, isometric shape. Their own electric and magnetic fields should not disturb metal deposition, and they should be plated with a uniform layer of nickel (Chapter 3). The nickel layer should be slightly thicker than the maximum linear diamond particle dimensions [218].

The main characteristic of electroplated (E. P.) tools is that there is normally only one layer of diamond particles. Consequently, such tools have a relatively large number of diamond points per unit cutting face. According to Brugger [26], the number of points on the cutting face of an E. P. tool may be several times higher compared with diamond impregnated metal bond tools. For example, Brugger calculated 1810 diamond points per cm^2 of E. P. tool surface area and only 410 points per cm^2 for a bronze bond tool of 100 concentration. The diamonds were 60/80 US mesh size. Another benefit is the low initial cost of E. P. tools, since no mould is needed. The main disadvantage is their comparatively short life, which comes to an end as soon as the diamond (or CBN) layer is worn in one place. The thickness of the nickel layer is approximately 1.5 times the diamond particle size. Moreover, surfaces machined with E. P. tools are comparatively rough. The use of nickel electroplated bonds is limited small tools, such as mounted wheels and points, and complicated profile tools.

Electroplating methods can be employed to bond diamonds not only to rigid supports, e.g. metal hubs including those with highly complex profiles, but also to flexible supports such as various kinds of fabric or wire mesh. Abrasive sheets made by co-depositing diamond powder and nickel are used for similar jobs as sheets based on sintered metal (e.g. for lapping), sheet thickness being constrained chiefly by the diamond particle size. In the wire mesh type of tool, the peak of each part of the mesh is left exposed on the weft and the warp and these tiny lands are then electroplated using diamond abrasives, the grit size of which is selected according to the application. Also in use are flexible tools with diamond-nickel deposited on fabric in the form of relatively coarse 'pellets' [138].

4.2.4 Resin bond tools

The main application of resin bonds is in wheels for grinding cemented (or sintered) carbides. The reason is the 'free cutting' or self-sharpening wheel behaviour, which results in lower specific grinding pressures and temperatures, as well as lower manufacturing costs compared to impregnated metal bond tools.

In most cases, resin bonds are made from phenolic resin powders, a characteristic property of which is the irreversible transition from a plastic to a hardened state. The change in state takes place as a result of cross-linking, i.e. the formation of a network of molecules which, in turn, results either from a reaction between a hardening agent and the resin powder molecules (the so-called chemi-setting plastics) or is a consequence of heating (the thermosetting plastics), the latter type of process being much more commonly employed in diamond tool manufacture. Some limited use is also made of thermoplastics.

Of the numerous resin bond types currently in use, the most popular are those based on variously modified phenolic resins. In quantitative terms, polyimide resins, amine resins etc. still play a marginal role. Bonds made of phenolic resin alone, without any additives, would however be too soft and abrade too quickly, quite apart from being susceptible to overheating. For these reasons, resin bonds are prepared from mixes of resin powder and what is known as a filler. For example, a resin bond as defined in Soviet Standard GOST 3552-63 comprises Novolac powder and at least 7% hexamine. The type and amount of filler can be varied to modify the properties of the diamond impregnated layer and, in this way, to improve tool performance with a given resin (Fig. 4.8). The additives may contribute from 40 to 75% of the total bond volume. For example, fine grit silicon carbide (SiC) or corundum (Al_2O_3) is added to increase tool life [214]. Special graphites or molybdenum disulphide are used to reduce frictional heat when grinding. Metal powders (Al, Ag) in the bond contribute to the rapid transfer of heat from the wheel/work-piece interface [182]. To obtain an especially soft action grinding wheel, cellulose particles are added [214]. Thus, resin bond properties can vary over a wide range.

Examples of bond compositions based on XYLOK 939P from

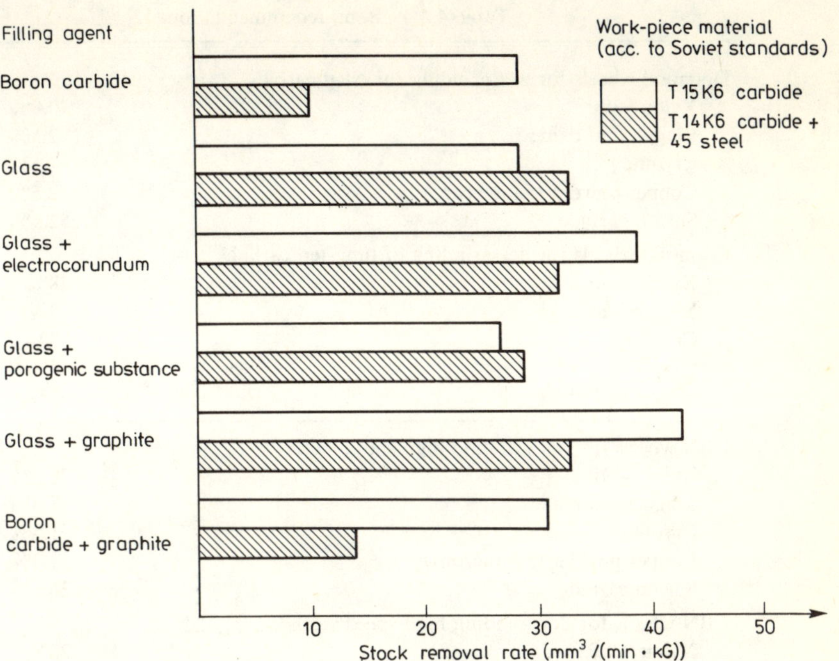

Fig. 4.8 — Stock removal rate for resin bond wheel versus the kind of filler in the bond [169].

Advanced Resins Ltd. in the UK for grinding cemented tungsten carbide are given in Table 4.4. XYLOK 939P is a member of the phenol-aralkyl family of high-performance thermosetting resins, suitable for the manufacture of diamond and CBN grinding wheels. Processing and curing parameters are very similar to those for phenolic resins, however XYLOK 939P offers superior heat stability when cured.

After mixing the components of a resin bond with the diamond or CBN abrasive, carefully weighed, the mixture is poured into a mould and an abrasive layer is produced by hot pressing. The temperature parameters of curing are in the 420–520 K range, and the pressures in the 30–80 MPa range. As the resin hardens, the diamond layer becomes bonded to the wheel hub or other tool body. Depending on the application, the hub or other support element may be made of steel, non-ferrous metals (e.g. aluminium alloys) or of plastic (often containing the resin employed as the bond material for the diamond or CBN powder). Metal hubs are made by machining and the area in which the superabrasive layer is to be set is milled to obtain a greater area of contact and/or better adhesion of the diamond layer to the hub.

The diamond abrasives best suited to resin bond tools are generally of the polycrystalline or aggregate type (Chapter 3). The performance of

Table 4.4 — Bond recommendations [2]

Diamond wheels for wet grinding tungsten carbide		
Xylok 939P	40.0%	(W/W)
Magnesium oxide	2.5%	,,
Cryolite	2.5%	,,
Copper powder (50 microns)	2.5%	,,
Silicon carbide	52.5%	,,
Diamond wheels for dry grinding of tungsten carbide		
Xylok 939P	28%	(W/W)
Magnesium oxide	3%	,,
Cryolite	3%	,,
Graphite	4%	,,
Silicon carbide	62%	,,
CBN wheels for wet grinding high speed steels		
Xylok 939P	70.0%	(W/W)
Magnesium oxide	1.3%	,,
Cryolite	1.3%	,,
Copper powder (50 microns)	1.3%	,,
Silicon carbide	26.1%	,,
CBN wheels for dry grinding high speed steels		
Xylok 939P	28%	(W/W)
Magnesium oxide	3%	,,
Cryolite	3%	,,
Graphite	4%	,,
Silicon carbide	62%	,,

such tools can be improved by using metal-clad diamond/CBN powder. A specific type of resin bond tool is a wheel with oriented needle-shaped diamonds. Designated CDA-L*, the diamonds are coated with a layer having magnetic properties. A method of orienting CDA-L particles in the matrix during wheel manufacture has been developed at the De Beers Diamond Research Laboratory, and it involves the setting up of an electromagnetic field surrounding the mould assembly [57]. The electromagnetic unit, which incorporates a core locked centrally through the mould, is designed to generate radial lines of electrical force with respect to the wheel. During the mould filling operation, the CDA-L particles — clad with a magnetically susceptible layer — align themselves parallel to the magnetic lines of force as they fall freely, with the bond mixture, through the force field (Fig. 4.9). This is followed by the normal pressing and curing procedure. Tools manufactured by this method are offered e.g. by Citco Inc. (USA) under the trade name Magna 1. Grain

* CDA-L diamond powder is no longer available — Translation Editor.

orientation and the allowances made for the anisotropy of diamond involve a 1.5–2 fold increase of the cutting points on the working face of the wheel which in practice translates into increased stock removal rate and improved surface finish [14, 105].

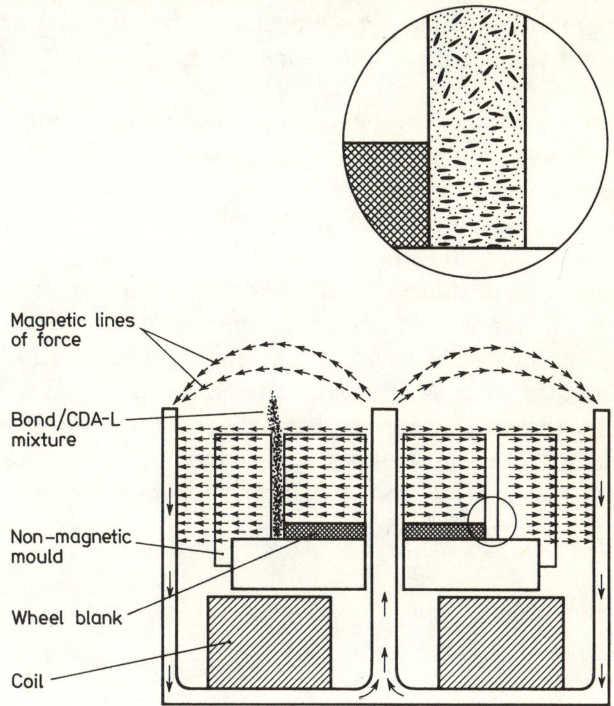

Fig. 4.9 — Schematic diagram of the electromagnetic orientation unit for the manufacture of 1A1 wheels [57].

4.2.5 Vitrified bond tools

Only a relatively small proportion of diamond impregnated products are made with vitrified bonds. A characteristic feature as compared with metal and resin bond products is their porosity, which may be as much as 55% of the diamond layer volume. By making room for the work-piece detritus, the pores play an extremely useful role, and at the same time facilitate distribution of the coolant/lubricant in the contact zone. A disadvantage of vitrified bond tools is their relative brittleness.

Vitrified bond tools are employed above all where materials with different properties that are in contact (e.g. carbide–non-ferrous metal–steel) have to be ground [10].

Two main techniques are used for the manufacture of vitrified bond diamond wheels or honing sticks:

— a shape corresponding to the tool dimensions is cold pressed from the abrasive mix, and is then freely sintered in a separate technological process;
— a hot pressing operation is employed similar to that employed for the manufacture of impregnated metal bond tools, where heating and pressing are conducted simultaneously. This method yields tools with improved hardness and wear resistance.

The main ingredients for the manufacture of vitrified bond diamond tools are powders obtained by milling fritted glasses and 'devitrificates' having softening temperatures from 900 to 1250 K. Good examples are boron glasses of the type $ZnO–B_2O_3–SiO_3$ or $Me_2O–Al_2O_3–B_2O_3–SiO_2$, where Me_2O stands for alkali metal oxides [10]. As a rule, the incorporation of substances causing crystallization of the glass (e.g. oxides of zirconium or titanium, or pure metals such as silver) increases mechanical strength. The bond composition may also include natural alumino-silicates such as feldspar and clay minerals. In view of the absence of sufficiently strong chemical type bonds at the diamond/ceramic interface, fillers in the form of high melting point abrasives (e.g. Al_2O_3) may be incorporated. It is also advantageous to use coated abrasive grains, especially ceramics containing unbound boron oxide [10].

5

Diamond abrasive products

Diamond products for abrasive machining, lapping or polishing applications can be classified into:

— tools in which diamond particles are bonded with metal, resin or ceramic compositions as described in Chapter 4;
— diamond coated (or electroplated) tools;
— lapping and polishing compounds and suspensions, including aerosol borne media.

Each of the above product types is highly specific in terms of structure, technology of production and uses.

5.1 DIAMOND IMPREGNATED TOOLS

Diamond impregnated tools represent the largest proportion of both synthetic and natural diamond use. The working part of a tool may be a single layer of diamonds, as in electroplated (E. P.) tools, or it may be an abrasive layer up to several mm thick (as in resin, metal or vitrified bond tools). The diamond content in the abrasive layer is called the diamond concentration. The parameter is an arbitrary one, but it is universally accepted. A 100 concentration means 4.4 ct of diamond per 1 cm^3 of diamond impregnated layer.

In most tool types the diamond layer is cemented to the tool blank in the course of the process of diamond sintering with the bond. In an alternative procedure use is made of previously prepared shapes (known as segments) or sheets with diamond, which are cemented to the tool blank by way of brazing, gluing, sintering or mechanical mounting (as in the manufacture of saws, milling cutters, etc.

The widespread use of abrasive tools has made it necessary to standardize the description of their structure and application characteristics. A typical designation, in the form of a letter-figure symbol, is made up of two parts, one of which refers to the tool design and size, and the other signifies the properties (composition) of the abrasive layer. International standards, e.g. FEPA, ISO, typically require the following design information to be specified:

(1) basic core shape,
(2) shape of diamond layer,
(3) location of diamond layer,
(4) modifications.

The application part of the tool designation should include such information as bond type, the type, size and concentration of diamond, etc. (Fig. 5.1). The major abrasive tool types and their applications are described below.

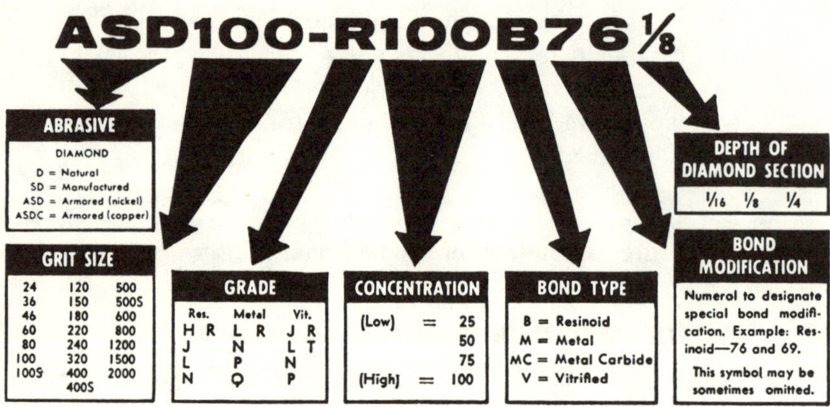

Fig. 5.1 — Specification of the diamond active layer in grinding wheels (Norton Co.) [142].

5.1.1 Diamond grinding wheels

One of the main uses of diamond abrasives is in grinding wheels. Table 5.1 lists the most common diamond wheel shapes and the kinds of operation to which they are appropriate. The general dependence of the grinding performance on the wheel characteristics and operating parameters is shown in Fig. 5.2. The stock (material) removal rate of a grinding wheel is defined as:

Table 5.1 — Diamond wheel applications [63]

Type	Form	Cylindrical grinding	Surface grinding	Internal grinding	Sawing	Sharpening of rake angle	Sharpening of clearance angle	Sharpening of lathe tools	Profiling
1A1	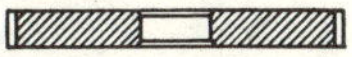	■	■	■				■	
1A1R					■				
1A1W				■					
1C2						■			
4A2	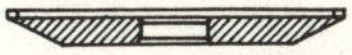					■			
4A9	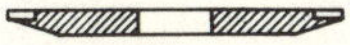								■
4B4	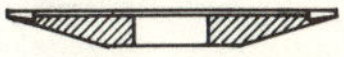								■
4B2	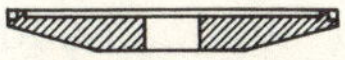					■			
4CH9	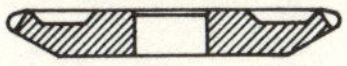					■			
4.11A2						■	■		
6A2	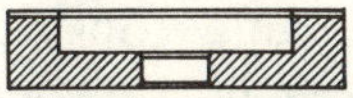		■				■	■	
6BA2	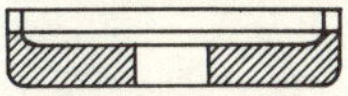						■		
9A3	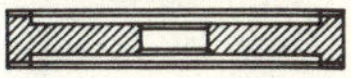							■	
11V9						■	■	■	
11VV5	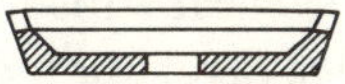						■		
12A2 20°	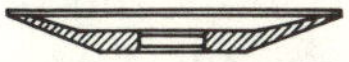					■	■		
12A2 45°	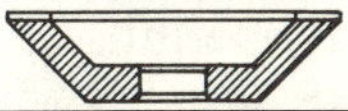		■			■	■	■	
12V9	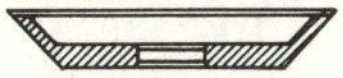					■	■		
14F1	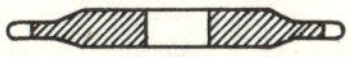								■

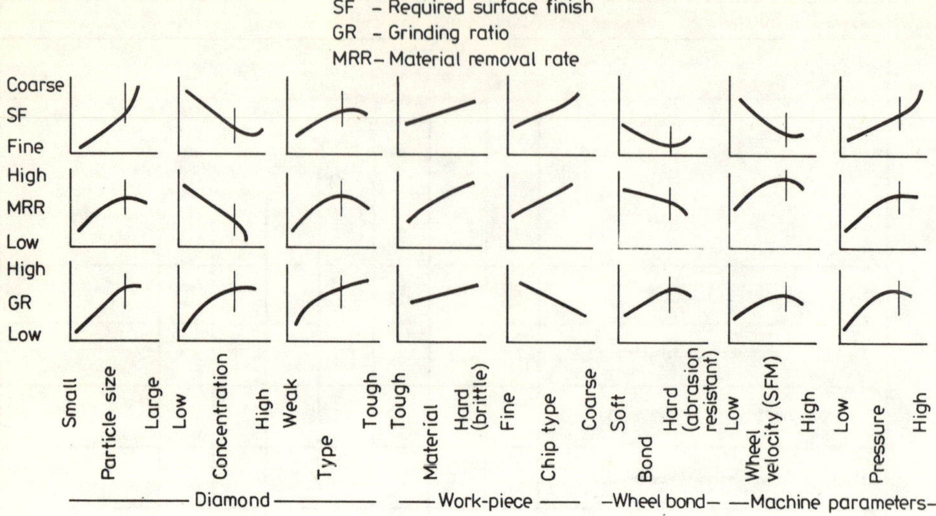

Fig. 5.2 — Effects of stock removal rate machining parameters and diamond content and particle size on wheel performance [86].

$$\text{MRR} = \frac{\text{Material removed}}{\text{Time}}, \quad \text{in cm}^3/\text{min}$$

The grinding ratio (G-ratio) describes the specific wear and, as such, the life of a diamond wheel under the set conditions. It is calculated thus:

$$\text{G-ratio} = \frac{\text{Material removed}}{\text{Wear of the diamond layer}}, \quad \text{in cm}^3/\text{cm}^3$$

As can be seen in Fig. 5.2, in selecting the most advantageous parameters defining grinding performance, account is taken of a wide range of variables.

One of the most important functional properties of a grinding wheel defining the range of its application is its size. Because of more favourable thermal and kinematic conditions, a grinding wheel is the more economical the greater its diameter. Due to the longer cooling phase the abrasive grain is also less susceptible to wear (this has a very positive effect on wheel life). As a rule, the abrasive layer should be used uniformly over its entire width. In mechanical grinding, this is achieved by the feed movement of the work-piece or of the grinding wheel. In cup and dish wheel grinding, the narrow abrasive layer permits quicker and cooler grinding. In offhand tool grinding with a support, a wider abrasive layer is better as it enables better control. In general, however, the greater the contact area, the narrower the abrasive layer width should be. If it is too wide, it will cause high frictional heat with negative consequences. Besides, an abrasive layer that is too wide can rarely be used to the full extent (Fig. 5.3). The abrasive particle size mainly determines the stock removal capacity with diamond, as well as the surface finish on the work-piece. Coarser particles make for better wheel life while finer particles improve the surface finish, but decrease the stock

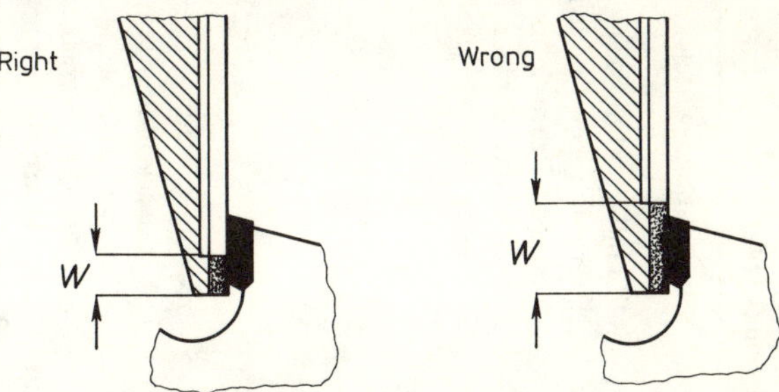

Fig. 5.3 — Selecting the width of the diamond layer relative to work-piece size.

removal rate. Depending on the contact conditions and the dimensions of the work-piece surface, the following diamond mesh sizes are recommended [157, 182, 207]:

— rough grinding — 60–120 US mesh,
— finish grinding — 140–230 US mesh,
— fine and precision grinding — 325 US mesh to 15 μm,
— mirror finish grinding — 10 to 2 μm.

High concentrations, above 75, are typical for coarse mesh diamond in peripheral wheels, profile wheels of small contact area and rim width [207]. In general, a higher concentration should be used for metal bonds. Low diamond concentrations of 25 to 50 are used with fine grits for wheels of large rim width, as well as for face grinding with cup wheels.

The recommended applications of diamond wheels with different bonds are listed in Table 5.2. Note that each bond type will have a range of hardness and abrasion resistance values. Generally speaking, hard bonds are associated with narrow abrasive layers, coarse mesh diamond, high concentrations, long tool life, good profile retention and wet grinding [63]. Softer bonds are generally more suitable for wide abrasive layers, fine mesh diamond, low concentrations, work-pieces sensitive to heat, dry grinding conditions. The grinding results may differ greatly from one machine to another. Therefore, the machine must meet the following basic requirements: rigid construction, powerful drive, accurate spindle mounting, table tracks free of play, smooth work-piece feed, vibration-free mounting. Efficient grinding performance with

Table 5.2 — Diamond wheel applications with different bonds [142]

Metal bond wheels	Resin bond wheels	Vitrified bond wheels
Offhand grinding, using cup wheels 1. Sharpening single-point carbide tools 2. Reconditioning excessively dull or chipped single-point carbide tools 3. Grinding new "milled and brazed" tools 4. Edging quartz crystals and synthetic gemstones	**Offhand grinding, using cup wheels** 1. Grinding new single-point carbide tools on a production basis 2. Sharpening dulled single-point tools where fast cutting action is a primary consideration	**Offhand grinding, using cup wheels** 1. Grinding new "milled and brazed" single-point cemented carbide tools 2. Sharpening single-point cemented carbide tools 3. Edging quartz crystals and synthetic gemstones
Cutting-off, slotting, and sawing 1. Cutting sintered carbide blanks to desired lengths 2. Cutting rough semi-sintered blanks to desired sizes and shapes 3. Salvaging chipped carbide tools by cutting-off the damaged portion of the carbide tip 4. Sawing glass 5. Sawing germanium, silicon, quartz into "wafers" 6. Sawing stone, such as marble, granite, slate, etc.	**Cutting-off and slotting** Cutting sintered carbide blanks to desired lengths where fast cutting action is essential	
Precision grinding 1. Generating curved surfaces on lenses 2. Surfacing, edging and beveling lenses 3. Surfacing and "milling" glass prisms 4. Grinding ceramic insulators, etc. 5. Pencil edging of glass 6. Bevelling of glass	**Precision grinding** 1. Cylindrical, centreless, surface and internal grinding of carbide tools, cutters, dies, gauges, etc. 2. Sharpening carbide inserted-tooth milling cutters, reamers, drills, etc. 3. Grinding threads from solid in carbide taps and gauges 4. Grinding chip breaker grooves in carbide tools 5. Internal grinding cartridge dies of special dense and wear resistant steels 6. Finish surfacing glass prisms, lenses, synthetic gemstones, etc.	**Precision grinding** 1. Surface grinding cemented carbide form tools, cutter blades, dies, gauges, mould liners, etc. 2. Grinding chip breaker grooves where wheels required are not thinner than 1/4" 3. Internal grinding tungsten carbide Vitrified bonded diamond wheels are available in many standard sizes in both straight wheels and plain cup wheels. A desirable combination of rapid stock removal and durability is characteristic of these wheels

minimum wear is possible only with wheels manufactured within the narrowest possible tolerances on balance and running truth (parallel and radial). Diamond wheels have to be manufactured very carefully, to ensure that the lateral run-out and the vertical eccentricity do not exceed 0.02 mm when they are correctly mounted on the spindle and suitable flanges are used. The wheel bores must be to H7 tolerance (H6 on special request).

Excessive axial and radial run-out of diamond wheels must be reduced and if possible eliminated by truing. Three basic truing procedures are in use (Fig. 5.4):

— with a brake truing device,
— on a mild steel block,
— with loose SiC.

In truing with a brake truing device, a separate truing unit on a sliding support carriage is equipped with a silicon carbide (SiC) wheel of grain size 70, grade M. Feed rates up to 0.05 mm can be set. The truing wheel should be rotated by hand before it touches the diamond wheel.

Truing on a mild steel block is in fact a grinding operation. The abrasive effect of long chip steel material on the wheel matrix results in rapid matrix erosion and gives a true running wheel within a short time, as well as a very open, free cutting rim surface.

In the third procedure, cup wheels with a wide abrasive layer can be trued using loose silicon carbide abrasive of grain size 80–140. The SiC powder is poured onto a glass or wear-resistant steel plate, and the diamond wheel is then moved over the plate using a light pressure.

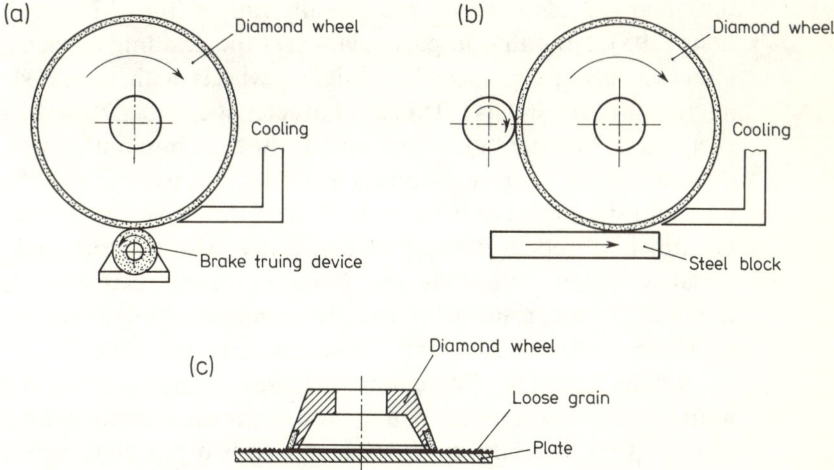

Fig. 5.4 — Modes of diamond wheel dressing: (a) truing with a brake truing device; (b) truing on a soft steel block; (c) truing with loose SiC.

After geometrical shaping (= truing) of diamond wheels, it is absolutely necessary to dress or prepare the surface of these wheels prior to use. According to recommendations adopted by Norton Co., diamond wheels should in general be dressed with 37C grain ceramic sticks—the SiC grain size employed should decrease with the diamond grit size. Thus, for example, with 60 US mesh diamond, the 37C abrasive grain size should be No. 30, while for 220 US mesh diamond, it should be No. 80.

As with other diamond abrasive products, diamond impregnated grinding wheels perform well on hard non-ferrous materials such as sintered (WC–Co) carbides, ceramics, glass, ferrites, natural and synthetic construction materials, stone, semiconductor materials, PCD and PCBN tool materials, etc.

The most typical methods of working carbides using diamond wheels include: sharpening of cutting tools, surface grinding, internal and cylindrical grinding, finish grinding after cutting-off or drilling operations, etc. When grinding carbides with diamond wheels, two sets of costs must be considered: labour and overhead cost per piece, and diamond wheel cost per piece. Labour and overhead cost per piece can be reduced by increasing the material removal rate. This increases the diamond wheel cost per piece, since wheel wear increases with the material removal rate (see Fig. 5.6). Generally, the economic material removal rate — i.e. the rate at which labour and overhead and wheel costs per piece are balanced for lowest total grinding cost—is quite low in comparison with grinding steel. For dry tool and cutter grinding, the optimum material removal rate (MRR) is seldom more than 16.4–32.8 cm^3 of carbide per hour. For wet surface, cylindrical and vertical spindle grinding, the optimum carbide removal rate usually ranges from 32.8–154 cm^3 per hour [195]. Note that in particular cases the grinding efficiency can be increased using e.g. specially designed wheels with needle shaped or aggregate particles [64]. Typical characteristics describing the grinding costs as a function of wheel speed show a minimum which shifts downwards as coarser diamond particles are used (Fig. 5.5). As the diamond size increases, so also does the G-ratio. Similarly, a maximum G-ratio is observed when plotted against wheel speed, and it shifts towards higher values as the bond hardness increases [86]. The parameters for carbide grinding with diamond wheels as recommended by Diamant Boart (Belgium) are summarized in Table 5.3.

A popular mode of machining sintered carbide, also used on other hard, multi-phase, current conducting materials is electrolytic grinding. This is very similar in principle to electro-polishing, metal being removed from the work-piece by electrolytic action using a rotary cathode, which gives a high rate of material removal and a good surface

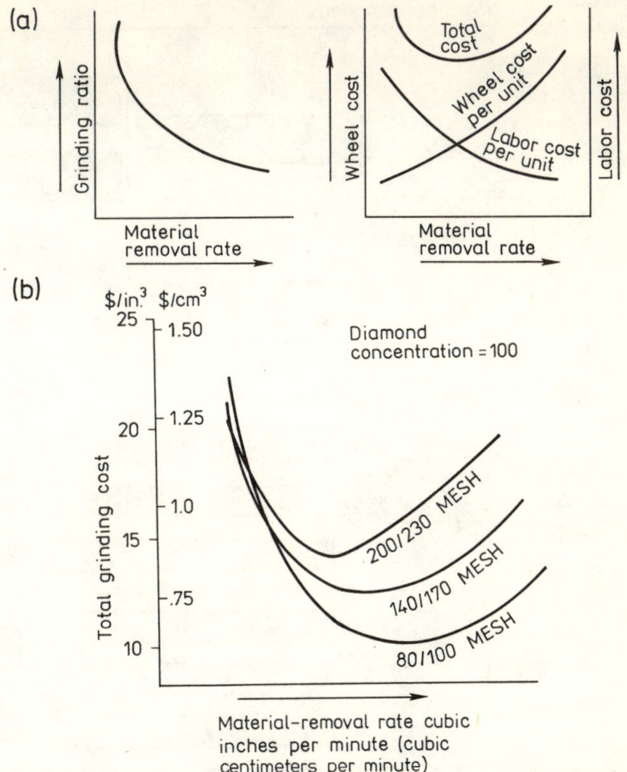

Fig. 5.5 — Effect of stock removal rate on grinding costs [195].

finish in one pass. The electrolyte is forced through a controlled gap of approximately 0.025 mm between the work-piece (anode) and the diamond particles above the wheel bond (cathode). A high density, low voltage direct current is then applied which causes disintegration of the Co skeleton binding the WC grains. Current densities are varied to suit the wheel/work-piece contact area (Fig. 5.6). Approximately 90% of the

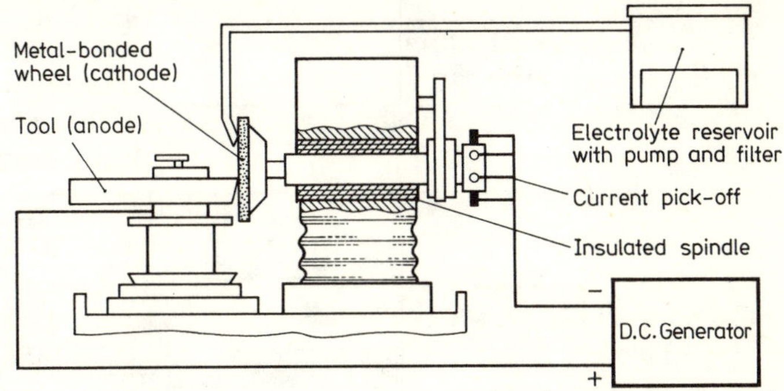

Fig. 5.6 — Schematic diagram of equipment for electrochemical grinding of sintered carbides with diamond wheels [62].

Table 5.3 — The parameters of using of diamond wheels

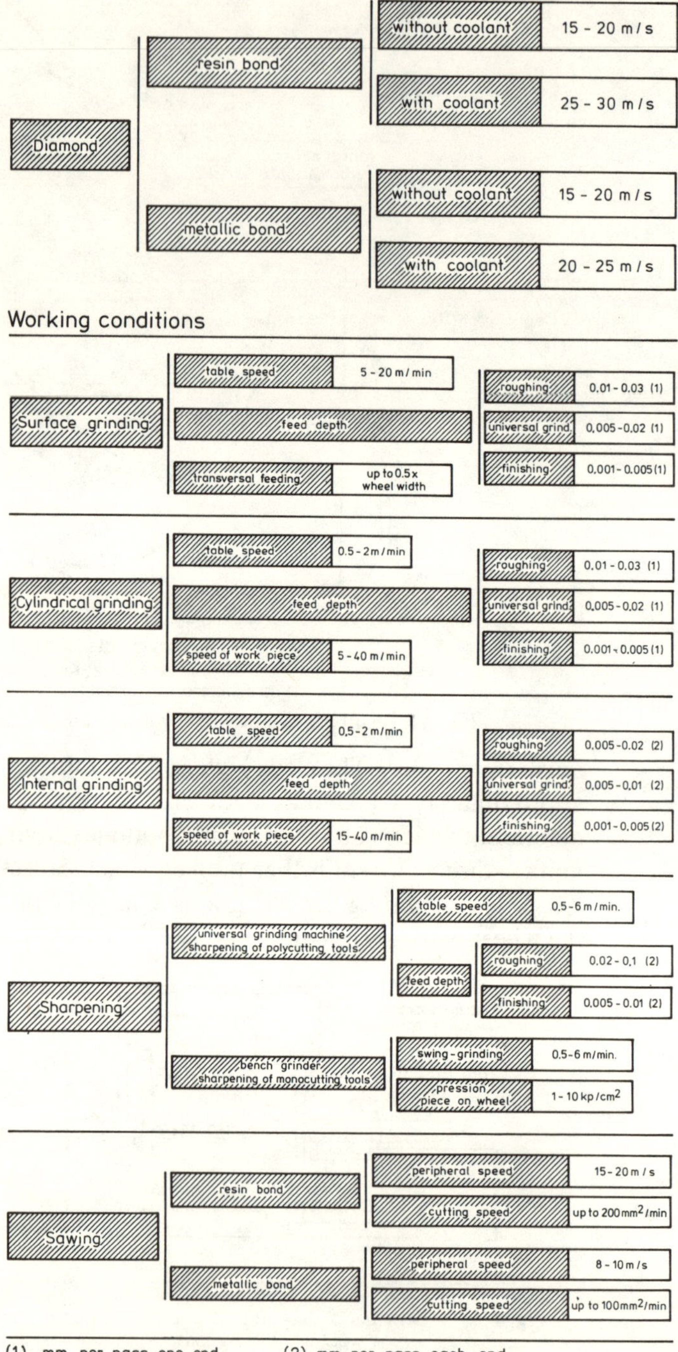

work-piece material is removed by electrolytic action, and the remaining 10% — by the diamond impregnated wheel. It will be obvious that owing to the very nature of the process, wear on the diamond wheel is very slight and it is estimated that the life is up to 20 times longer than in conventional mechanical grinding [62, 82].

The electrolyte is generally a conductive solution of sodium nitrate, disodium hydrogen phosphate and sodium bicarbonate [82]. Special machine tools and metal bond diamond wheels are of course required for electrolytic grinding.

In Table 5.4 are listed the most common glass working operations carried out using diamond abrasive tools. Rough grinding is performed with 200–125 μm size diamond abrasive, while finish grinding is

Table 5.4 — Typical applications of diamond tools in the glass industry [62]

Material	Application	Material	Application
Quartz crystal production	Heavy duty block crystal cutting	Plate glass	Automatic pencil edging
	Automatic wafer cutting		Automatic bevelling
	Lens generation (spherical)		Automatic arrising
	Surface and cylindrical grinding		Automatic flat edging and mitreing
	Heavy duty surface grinding (vertical spindle machines)		Flextol bevelling
Ophthalmic lens production optical glass	Lens generation (spherical)		Flextol bevelling (hand)
	Toric lens generation Tecnaphot 122 machines Autoflow radmaster		Flextol pencil edging
	Automatic lens edging		Pencil edging (hand)
	Hand edging		Brilliant cutting
	Fused bi-focal trepanning		Slotting saws (flexible drive machines
	Lens slotting		Automatic circular bevelling
			Heavy duty cutting and slitting
General glass grinding applications	Thin wall glass tube cutting	Scientific lens production optical glass grinding	Lens generation (spherical)
	End grinding tubes		Heavy duty lens grinding
	Taper grinding ground glass joints male and female		Heavy duty flat grinding
	Internal grinding		Automatic edging and bevelling
	Honing hypodermic tubes		Surface and cylindrical grinding (horizontal spindle)
	Glass tumbler grinding		Lens blank production by trepanning
	Decorative glass tube cutting		

generally done with 50–28 μm size particles. In special cases still finer micron diamond or non-diamond materials may be used.

Of particular importance is the use of diamond tools in optical lens manufacture. The first operation, generation of the basic curve on a plain lens blank, is carried out on semiautomatic or automatic machines using a special type of diamond impregnated cup wheel. This is followed by smoothing. Two types of tool are used for this operation: diamond impregnated pellets stuck on to the surface of a metal holder of appropriate radius, and solid type diamond tools entirely covered with diamond impregnation (see Fig. 5.8). The majority of the pellets are 6–10 mm in diameter. Compared with the solid type tools, there is a greater flexibility of use since the same pellets can be stuck on to supports of different radii; they also allow the glass to be more easily removed. Centring and edging wheels are used for grinding, finishing and chamfering lens edges. These tools usually work on their periphery under constant pressure, the lens being rotated at a speed which varies according to the depth of cut. In the case of a spectacle lens, the rims are shaped to the size of frame in which it is to be mounted.

An important application of diamond abrasive is in crystal glass grinding, or decorative cutting. Metal bond diamond wheels with a variety of profiles are used: flat, angular, radial (see Fig. 5.9). The peripheral wheel speed should be 25–40 m/s [63]. Water is used as a coolant with a small amount (2–3%) of oil added. The flow rate must be high, and the jet must fall precisely in the area of wheel/glass contact. The wheels require periodic dressing to maintain a free-cutting action and to eliminate profile deformations. Also widely used are dish wheels for grinding chamfers on door and mirror glass, and so-called pencil edging wheels for rounding the edge of motor car windows on automatic grinders [94]. Another example of rounding and smoothing of sharp edges is the 'flatting' of drinking glass rims.

Owing to the large variations in grindability as well as the shapes and sizes of ceramic parts, not to mention the varying surface finish requirements, a wide range of diamond wheel types is employed in the ceramics industry. Dimensional tolerances, as well as those related to plane parallelism, roundness or surface finish of ceramic components depend on the intended application. The requirements are fairly liberal in the case of refractory and sanitary ware, but very strict for products for electronics applications where dimensional accuracy may be measured in nanometres (nm). One field in which ultrahigh precision is absolutely essential is in the grinding of piezoelectric materials such as lead zirconate or barium titanate which are used in underwater hydrophones and sonars, since the quality of surface finish determines electrical frequencies. Illustrative of the close tolerances achievable with diamond wheels, ceramic substrates can be ground down to 5 μm thick for laboratory testing [209].

The most widely used mechanical machining operations on ceramic parts are surface and cylindrical grinding. As a rule, diamond tools are metal bonded except for fine finishing operations when resin bonds are used. Some ceramics are so brittle that the softer resin bond is chosen for its cushioning effect [209]. One example is in centreless grinding, when a rubber control wheel frequently partners a wide resin bond diamond wheel to ensure a shock-free 'ride' for free-rolling cylindrical components such as minute aluminium oxide fuse insulator tubes.

Fig. 5.7 — Examples of diamond abrasive tools for stone working (courtesy of Diamant Boart SA, Belgium).

Stone surfacing embraces a wide range of operations which, following sawing or milling, grinds to near final thickness and gives the surface its final polish. The tools most commonly employed include flexible discs, rigid backing plates with removable segments, a head with 'satellite' plates, surfacing wheels for portable machines, wheels for grinding edges, etc. (Fig. 5.7). The choice of tool depends on the stone to be worked, the surface shape, and the machine type.

5.1.2 Diamond abrasive profiling tools

The shape of a diamond abrasive profiling tool is a mirror image of the surface to be worked. A highly complex profile can be attained on stone components using one cutter of the required shape or a series of cylindrical elements. The contouring and finishing of highly complex surfaces is often effected with electroplated tools [63]. Simple cutters have a segmented structure, with the segments mounted in one plane on the lower (working) part of the tool. Sometimes the segments are mounted at the front end of the cutter or near the flange. To facilitate removal of the detritus and to increase the area of tool/work-piece contact when large plane surfaces are involved, the diamond segments are often rounded. Plane surfaces can also be machined with cylindrical cutters. Such cutters are especially useful on narrow concave or convex surfaces. A typical tool design is shown in Fig. 5.8.

After sawing, stone blocks or slabs are generally subjected to milling, grinding (calibrating) and polishing. Milling imparts planeity or the profile and surface finish required for subsequent surfacing operations. Milling cutter designs include one-piece face milling heads, milling heads with indexable diamond segments, rolls, diamond cup wheels, profile grinding wheels, etc. Thus almost any shape, even quite complex, can be imparted to stone products (Fig. 5.7).

Diamond tools are also used in machining stone tiles for new buildings, but they are also widely employed in the reconstruction and repair of old, historic buildings. The shapes of milling cutters can be programmed to obtain profiles resembling the manual work of stone masons. The profiles include all kinds of cornices, stair risers and treads, pillars, pilasters, etc.

A relatively wide variety of tools are used for shaping glass products. The purpose is usually to shape the glass edges so as to obtain a uniform profile free from sharp angles. The shape of the profile to be ground essentially depends on the intended use of the glass pieces. The types of edging correspond to the main applications of flat glass. The profiles most frequently found in this kind of work are of four types [63]: pencil edging — car windows, furniture glass, etc.; flat edging — show

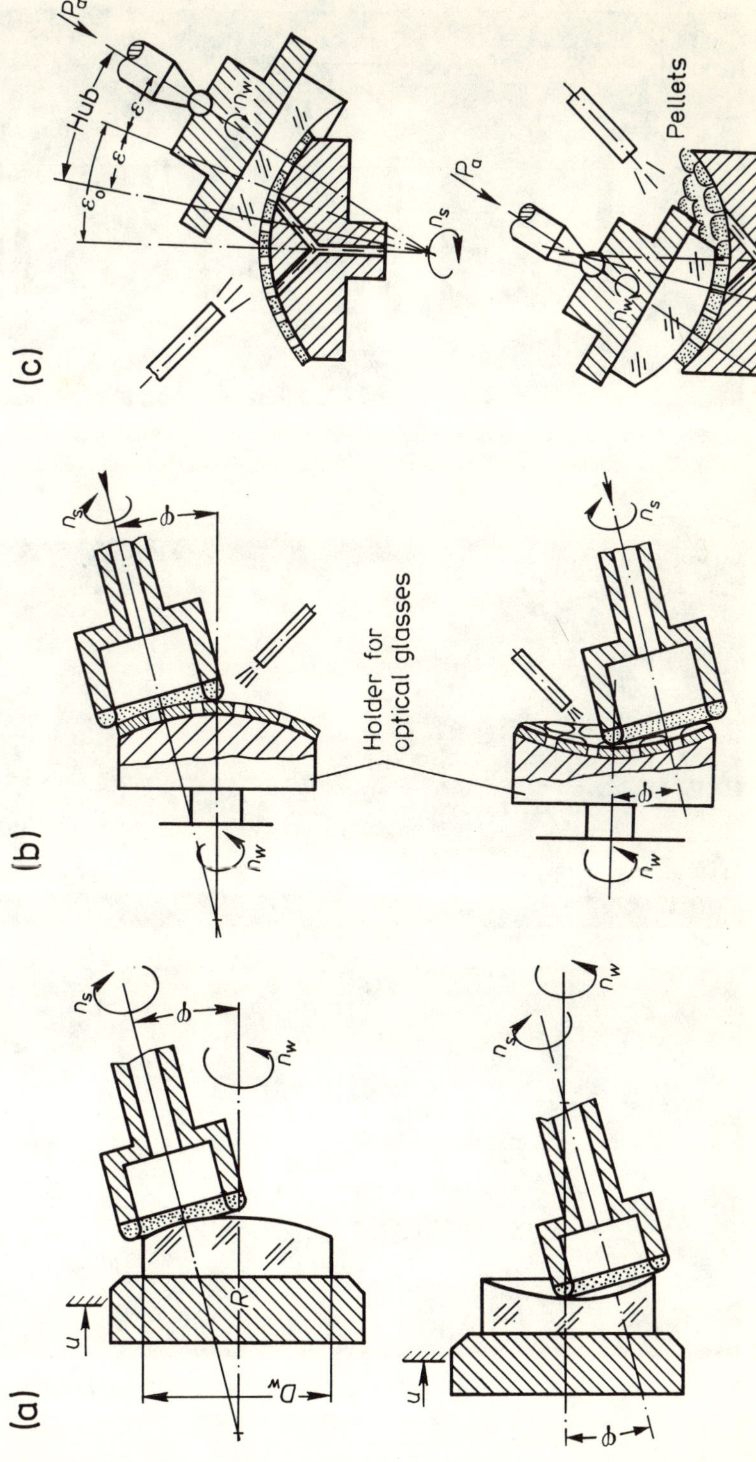

Fig. 5.8 — Diamond cutters in lens manufacture (a, b), and the use of pellets in lens grinding (c) [218].

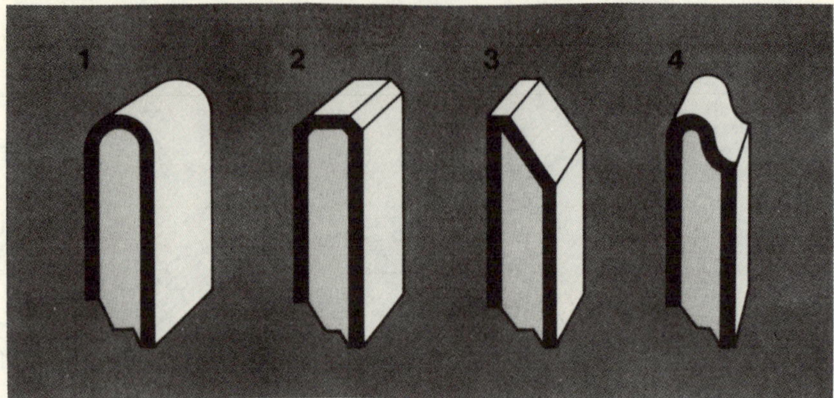

Fig. 5.9 — The shape of diamond wheels for the edging of flat glass: *1* — pencil edging, *2* — flat edging, *3* — bevelling, *4* — OG profile [63].

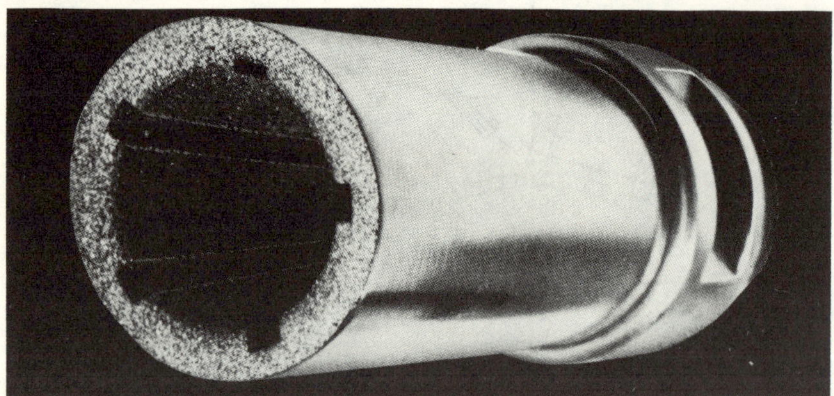

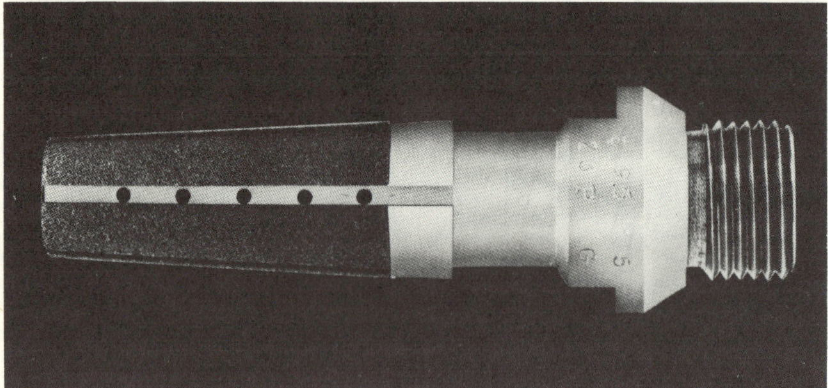

Fig. 5.10 — Conical reamers for grinding conical glass connectors (courtesy of Diamant Boart SA, Belgium).

windows, glass for furniture, doors, slabs, etc.; bevelling — mirrors, driving mirrors, etc.; OG-profile — table tops (Fig. 5.9).

Diamond tools are also used in the manufacture of conical connectors for glass laboratory equipment, and stoppers for glass containers (e.g. flasks). Male and female conical diamond electroplated reamers are employed (Fig. 5.10). The diamond reamers are mounted in drill chucks provided with coolant supply.

5.1.3 Diamond drill tools

Diamond drill bits embrace a class of tools for making round holes. The most common designs among diamond abrasive tools are:

- core drilling tools which have diamond segments at one end of a core barrel so that the barrel can be mounted in a machine which imparts a rotating motion to it;
- drill bits in the form of solid pins electroplated with diamond, which work along their axes of rotation and/or by performing an oscillating motion. In this class one may also include twist drills, the flutes of which are covered with an electroplated diamond layer.

Core drills are used not only for making holes but also for cutting out cylindrical samples from various hard engineering materials. Note that the tools designed for sampling the earth's crust, discussed in Section 6.5, are traditionally regarded as a separate class.

In terms of their design and diamond layer thickness, drill tools are classified into:

- continuous crown drills,
- segmented crown drills,
- segmented core drills,
- coring bits.

Each of the four classes is characterized in Table 5.5. Continuous crown drills have smaller diameters than segmented crown drills. As the tool diameter increases so may its wall thickness, to a maximum of 3–5 mm. The diamonds may be set in the working front of the drill or they may form an abrasive element made up of fine diamond grain. Surface-set drill bits give good penetration rates and are preferred when drilling lightly reinforced concrete containing abrasive aggregates such as sandstones, limestones and basalts. They are not recommended for penetrating heavily reinforced concrete which contains quartzite or flint aggregates. Surface-set tools are prone to damage through misuse. Much more frequently used are impregnated bits which usually give slightly

Table 5.5 — Diamond core drills for the construction industry [63]

Continuous rim drill bits The active part of the tool consists of a diamond-impregnated crown with waterways to facilitate the evacuation of cooling fluid and cuttings. Continuous crown drills are widely used for drilling small diameters	
Segmented rim drill bits The active part of the tool consists of diamond segments. These are brazed onto the tool body so as to leave sufficient space between them for the evacuation of cooling fluid and cuttings	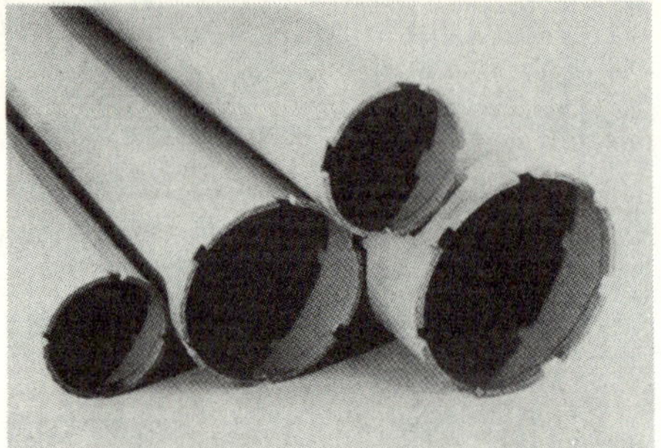
Segmented core bits The active part of the tool consists of diamond segments. These are brazed onto the tool body so as to leave sufficient space between them for the evacuation of cooling fluid and cuttings. Core drills are characterized by a more rugged fitting type	
Core bits Coring bits, adapted for civil engineering applications, are the solution for deep hole drilling. Unless otherwise specified when ordering, coring bits are manufactured with a male Craelius fitting. Coring bits may also be equipped with a core extracting device	

slower penetration rates but have longer lives. An important feature is that the hardness of the matrix can be varied to suit the application. For cutting relatively soft, abrasive materials a hard matrix is selected giving long tool life, but such a bit would only glaze when used on a hard stone. Therefore, in the latter case, the matrix is designed to erode away more easily, bringing fresh diamond points to the surface as drilling progresses. According to J. K. Smit and Co. [186], drilling efficiency depends on:

— the bit design, diamond type, size and concentration, and the matrix formulation,
— properties of the material being drilled, including its structural homogeneity, content of hard-wearing crystalline phases, reinforcement with other materials (e.g. steel),
— rigidity of the drilling machine,
— power of the spindle motor,
— the strength and diligence of the machine operator.

The importance of diamond tools for sawing and drilling construction materials is underscored by the fact that in the USA today, such uses account for the consumption of over 2/3 of all high mechanical strength diamond grit [145]. The use of diamond tools invariably saves time. Other reasons for their popularity include high precision, lower risk of structural damage (e.g. due to vibrations when using impact methods), lower labour consumption, lower operator fatigue. The old fashioned methods for making openings in concrete or brick walls involved the use of either a jackhammer, a sledgehammer, ball and chain or other demolition tool. In many instances these techniques also have some obvious *economic disadvantages* compared with diamond sawing and drilling. Besides the economic advantages of diamond tools for controlled demolition, they are also usually much quieter, which is consistent with present-day environmental requirements [218].

The diamond core drill is utilized in a very wide variety of applications: for the installation of electrical conduits, plumbing, sprinkler systems, material strength test sampling, for door or window openings, etc.

In the stone quarry diamond core drills are used to make holes for the insertion of diamond wire, or in the stone works to make holes in facing slabs. Such holes are necessary e.g. for screws with which to mount the slabs on the walls of buildings, or they may be useful in decorative work (arrangements involving various materials). The diameter of thin-wall core drills used in the stone industry ranges from a few mm to almost 800 mm. Also in use are multiple diameter drills that are employed for cutting out circles with holes in them, e.g. sandstone rings for flour milling or knife sharpening.

In building construction site situations it is often impossible or impractical to provide a continuous coolant supply. In such cases, use may be made of a recirculating coolant system (Fig. 5.11). Air mist or even dry drilling are further possibilities.

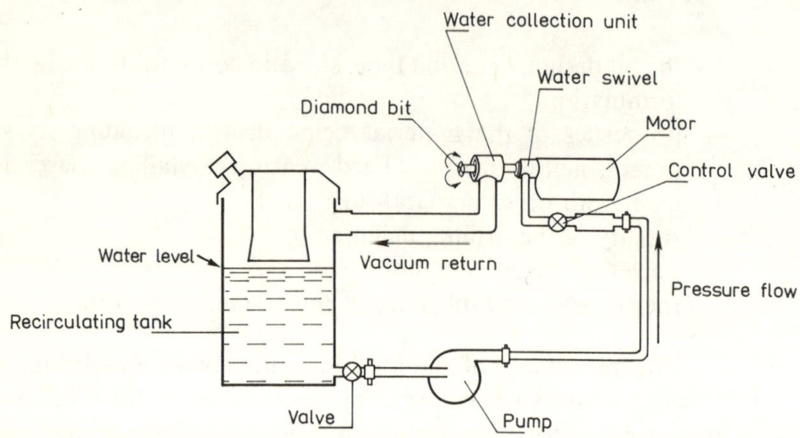

Fig. 5.11 — Schematic diagram of recirculating coolant system employed when making holes with a thin-walled diamond drill (e.g. in walls of buildings) [37].

An example of drill tools for ceramics working are the diamond 'probes' used in ultrasonic machining. The diamond coated surface reduces tool wear to a minimum, enabling extremely close tolerances to be maintained [39].

In semiconducting materials, holes are made with the wafers mounted on a hard support (glass or ceramic) to avoid chipping of the hole on the exit side.

Factors which affect the results of glass drilling include drill peripheral speed, wall thickness and the tool design, contact pressure and the coolant flow rate. Depending on the drill diameter, the following r.p.m.'s are recommended [213] : 8000–6000 r.p.m. for diameters of 2–5 mm ; 4500–3000 r.p.m. for diameters of 11–20 mm ; and 3000–1500 r.p.m. for diameters of 21–40 mm. Peripheral speeds vary in the range 1.5–2.5 m/s. The contact pressure should be adjusted so as to produce a penetration rate of up to 3–4 cm/min. Excessive pressure may not only cause premature drill bit wear but also break the slab being drilled. Coolant pressure should be adjusted depending on drill bit diameter. For 2–5 mm diameter bits the recommended [213] pressure is up to 50 N/cm^2, while for 21–40 mm diameter bits it is 5–10 N/cm^2.

Diamond electroplated ‘pins’ may be used for low stock removal machine or manual drilling, e.g. in art engraving. To dispose of the stock removed, lateral movements are made with the drill bit which inevitably gives. the hole a conical, barrel-like etc. shape. A true cylindrical hole is thus impossible to attain in this way. Electroplated drill bits may come in a variety of shapes (Fig. 5.12). They are widely used e.g. in dental care and for decorative engraving.

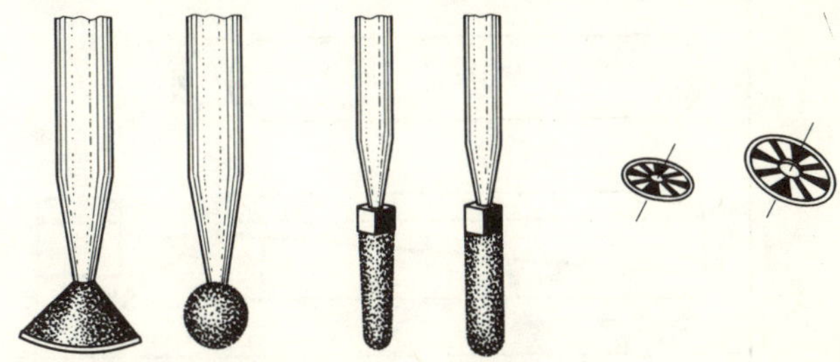

Fig. 5.12 — Selected diamond drills used in dentistry and in engraving.

5.1.4 Diamond reaming tools and hones

One type of abrasive tool employed in bore machining is the diamond reamer which is capable of producing geometrically accurate bores with a controlled surface finish in a set time cycle. A good application example is in the finishing of cast iron valves for controlling the direction and flow rate of liquids used in high pressure hydraulic apparatus. Diamond reamers are also used in the finishing of bores in sintered carbides and other hard materials. The difficulties typically encountered in honing parts with wide keyways, or with lands as in gudgeon pin bores, are now being overcome by diamond reaming techniques [62].

In their design, diamond reamers resemble conical grinding pins or sleeves. In one design, the centre portion is cone shaped with a reamer in the form of a screw-type adapter sleeve mounted on it (e.g. the DIAEXPANDER range from Diagrit, UK). The sleeve is capable of expanding within narrow limits which permits a tight fit on the pin. At the same time, the ability of the sleeve to slide along the pin makes it possible to increase or decrease its O. D. and thus to finish bores to predetermined tolerances.

Honing tools are designed above all for smoothing bores in any material, in the course of which process the tool makes a rotational and a rectilinear-reciprocating motion while the work-piece remains motionless. Conversely, the tool may be set in rotary motion while the work-piece performs a rectilinear-reciprocating motion. Such a procedure has the effect not only of reducing the roughness of the work-piece surface, but

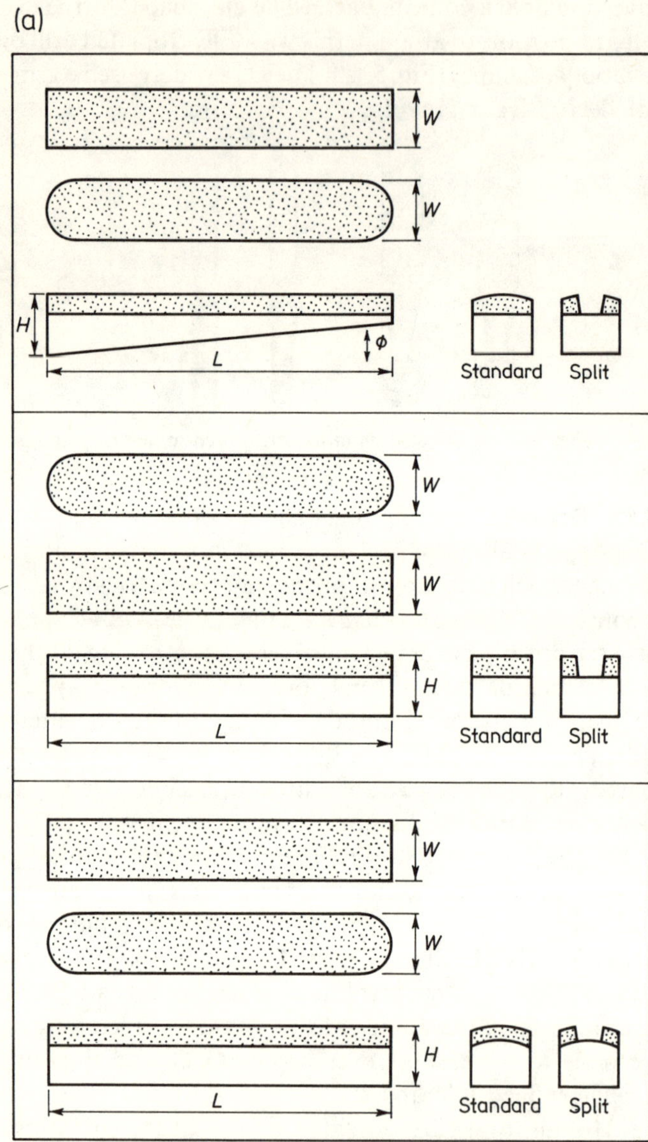

Fig. 5.13 — Typical diamond segments used in hones (a),

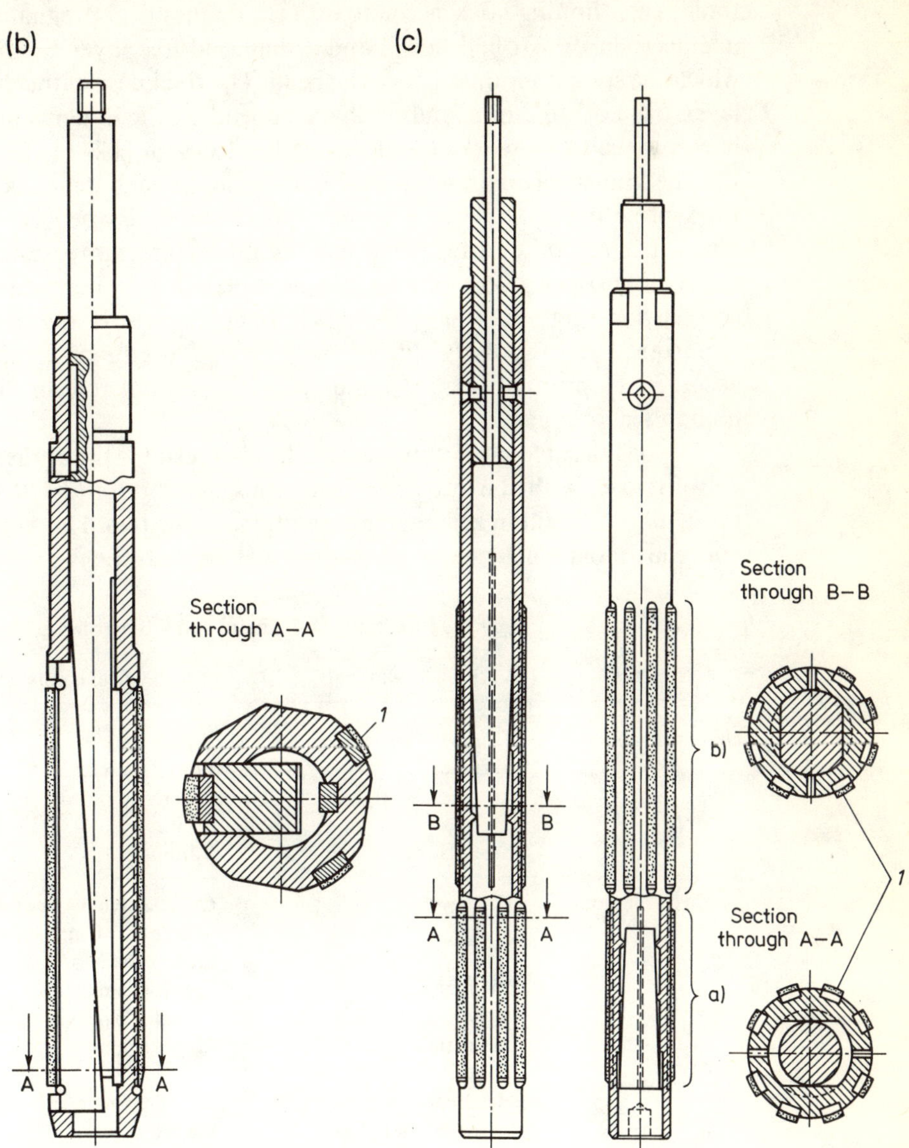

and ways of mounting them (b, c).

also of partially correcting its geometry. A purpose-designed machine tool is used. The honing head has sets of honing sticks mounted on it (Fig. 5.13). The design of the honing head resembles that of segmented tools; each honing stick is made up of a diamond impregnated layer (metal, resin or vitrified bond) and a diamond-free layer by means of which the stick is mounted onto the head. The thickness of the diamond layer, from 0.5 to 2 mm, and its shape depend on the dimensions of the bore surface to be worked and on the head design [62, 112].

The choice of matrix for diamond honing sticks depends on the work-piece material. Thus a copper-based matrix is used for honing bores in glass or quartz, while ferrous bonds are more suitable for heavy-duty operations on e.g. chrome-plated cylinder sleeves for combustion engines, where the stock removal rate is relatively less important than the useful life of the tool and the finish quality and accuracy. Resin and vitrified bonds play a marginal role in diamond honing technology.

Of considerable importance for the honing result is the abrasive type and grit size, with the superhard materials diamond and CBN being much more efficient in general than traditional abrasives. The work-piece materials most efficiently machined with diamond hones are listed in

Table 5.6 — Typical applications of diamond and CBN hones [112]

Abrasive	Work-piece material	Typical components
Natural diamond	Cemented carbide	Weapon parts
	Soft steel	Bolts and screws
		Connecting rods for 4-stroke engines
	Steel, hardened and soft	Various parts
Synthetic diamond	Cast iron	Cylinder blocks, motor stators, hydraulic control housings, brake cylinders
	Nitrided steel	Piston drums
	Hypereutectic aluminium	Cylinder blocks
	NF-metals	Piston drums
	Nicasil	
	Hard chromium layers	Cylinder liners
	Ceramics and glass	Electrical components
Cubic boron nitride	Hardened steel (62–64 HRC)	Connecting rods for 2-stroke engines
	Case-hardened steel	Gear wheels
	Cr-steels (60 HRC)	Injection pumps, bearings, hydraulic valves

Table 5.6. Probably the most widespread application of honing is in finishing cast iron cylinder bores in automobile engines. Engine bores are honed on highly automated machines with automatic size control and automatic inspection.

5.1.5 Diamond saw blades

In terms of design, saw blades fall into the following classes:

— circular saw blades,
— I. D. or annular saw blades,
— frame saw blades,
— band saw blades.

Circular saw blades show the greatest diversity in design, sizes and production technologies. The working layer of diamond may be affixed to the inside or more commonly the outside edge of the saw blade blank or centre.

Saws are tools in which diamond abrasive may be used in:

— segments brazed or laser welded to the tool blank;
— a layer electroplated onto the blank;
— a layer sintered directly onto the blank;
— a layer rolled or hammered into the tool periphery;
— a resin bond layer attached to the tool blank.

There are, in addition, two other groups of abrasive sawing tools intended for highly specialized applications, viz. dicing blades, used for cutting semiconductor materials, and diamond wire, used in quarries and stoneyards.

5.1.5.1 Circular saw blades

Circular saw blades are an extreme case of 1A type wheels with the diamond rim being either continuous (1A1 type) or slotted (1A1RSS type). Apart from that, circular blades come in a variety of slot sizes and shapes, and they may also vary in respect of mounting holes. Continuous diamond rims may be perforated and they may have variable thicknesses. The recommended applications for basic saw blade designs are listed in Table 5.7. Circular saw blades may range from a few mm up to 4 m diameter.

The primary application of circular diamond saw blades is in the cutting of stone, concrete and other construction materials. Such blades are strong enough for high stability automatic cutting as well as manual

Table 5.7 — Diamant Boart recommendations for the use

Blade	Type	Description	Dimensions
	J.C. Type	Continuous rim blade The active portion consists of a diamond impregnated ring of standard height 5 mm This ring is mounted on a special high strength steel support of a thickness inferior to that of the diamond impregnated ring so as to give a clearance during sawing	Diameters: from 200 to 400 mm Depth of the diamond layer: from 1.8 to 3.2 mm according to diameter
	B. Type	Narrow slot blade The diamond impregnated segments are inserted on the periphery of a slotted plate The distance between the teeth is reduced as compared with that of the standard type The number of teeth is higher than that of type D	Diameters: from 200 to 700 mm Segment thickness: from 1.8 to 5 mm according to diameter
	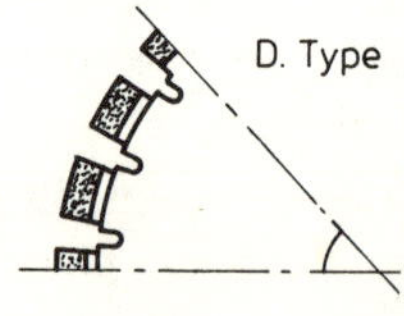 D. Type	Standard slot blade The active portion consists of diamond impregnated segments inserted on the periphery of a slotted plate	Diameters: from 200 to 3000 mm Segment thickness identical to that of type B: from 1.8 to 11.5 mm according to diameter

of different diamond saw blade designs [63]

General applications	Materials	Advantages	Recommendations
Recommended for all precise and delicate sawing Thin or small sized parts, and when an exceptional finish or a small loss of material is required	Any material of suitable dimensions Specially: enamelled earthenware, glass, brittle materials or with a fragile coating, etc. In some cases: polished marble	Outstanding finish and especially freedom from chipping Possibility of cutting small, brittle pieces Low kerf losses due to the reduced thickness of the tool	Mounting requires special care as the manufacturing tolerances are very strict Generous cooling Rotational speeds within 20 to 30 m/s Preferably hand sensitive feed Do not force the blade and make sure that it is correctly dressed (protruding diamonds)
Suitable for any sawing for which a good finish rather than a high cutting speed is required Used for cutting slabs of medium thickness and in particular polished slabs (granite, marble, etc.) Edge chipping is avoided when correctly used	In particular, polished slices Any material for which the finish is more important than the cutting speed	Due to the small space between segments, their number and narrow manufacturing tolerances, this blade design is ideal for sawing where a good finish is required with as little chipping as possible	Mount with special care: Observe the rotational direction indicated on the blade Generous cooling Sawing by successive passes for deep cuts in hard materials Hand or automatic operation Rotational speed: from 20 to 60 m/s The lower speeds are for hard and very compact materials, and the higher ones for abrasive and soft materials
Intented for rapid but good quality sawing Can be used for deep cutting Ideal blade for general production work	All stone and construction materials such as slate, concrete, granite, marble, bricks, tiles refractories, asbestos cement, etc.	High cutting speed and very satisfactory finish Capable of deep sawing Very low cutting cost	Mount with care Observe the rotational direction, generous cooling Sawing in one or several passes Hand or automatic operation Rotational speed: from 20 to 60 m/s depending on the material

operation. Not infrequently a number of saw blades are ganged together to form a cutting assembly e.g. for grinding and/or grooving of motorway surfaces. The noise when sawing is often a problem, but it can be reduced by affixing rings to the lateral blade surfaces to damp the noise-producing vibrations. One relevant design is the "Quiet Cut" device developed by Norton Co. [142].

The design of the saw blade blank, the thickness of the diamond rim and the way it is mounted onto the blade blank or centre all determine the width of cut as well as its rectilinearity and parallelism. If the blade blank is excessively thin, a wavy cut may result as well as adversely affecting the reproducibility of the cut quality. On the other hand, if the blade blank is too wide, the cut is also wide. When thin and parallel cuts are required, it is advisable to use saw blades of a different design.

According to O'Brien [145], three basic machines are utilized in the construction industry for concrete sawing and drilling: the wall saw, the floor saw and the core drill. All three can be powered by various sources: electric, pneumatic, gasoline (petrol) or hydraulic. The gasoline or petrol engine is by far the most common power source for a floor saw. A wall saw may be driven by pneumatic, high cycle electric or hydraulic systems.

Saws are used primarily for cutting conduits in the walls and floors of buildings, and for opening road surfaces e.g. to carry out repairs to the water or sewage pipes running below. They are also employed for chasing or cutting slots in concrete, asphalt, brick, stone and similar surfaces. Saws are operated manually or semi-automatically. Wall saws have many varied uses, with sawing out wall sections or door or window openings being the most common applications. Floor saws are utilized in a variety of horizontal surface applications: for sawing expansion joints, removal of deteriorated concrete sections, drainage gulleys, installing traffic control loops, bridge demolition, etc. [31, 46, 47, 108, 110]. A particularly common application of circular diamond saw blades is in the grooving of road surfaces to prevent hydroplaning (or aquaplaning) and to provide directional stability to motor vehicles. The grooves cut in roadways with diamond blades are generally 2.5 mm wide × 5 mm deep on 20 mm centres. Longitudinal grooves help to drain rain water from the road and provide resistance to sideways motion of the vehicle. Airport runways are generally grooved transversely to reduce the braking distance in wet weather. The texturing of concrete and asphalt surfaces to drain away water or other liquids is also widely employed in industrial and farm buildings. In road surface texturing applications diamond saw blades are used in sets of up to several dozen or even several hundred at a time (Table 5.8).

Table 5.8 — Typical diamond tools used in the stone processing industry [63]

Operation	Type of work or product	Type of tools
Quarrying	Chequering of the quarry floor	Large diameter blades
		Diamond chain
		Diamond wire
Sawing	Squaring of blocks	Gang saw blades
	Slabs	Large diameter blades
		Diamond wire
Cutting	Plates	Standard circular saws
	Tiles	SW silent type circular saws
	Strips	
Milling	Trimming to size	Face mills
	Trimming to thickness	Side mills
		Peripheral segmented wheels
		Rollers
Surfacing	Roughing	Satellite plates
	Smoothing	Flexible plates
	Prepolishing	Mills for edge polishing
		Wheels for portable machines
Profiling	Grooving and/or moulding in varying types	Peripheral wheels
		Cylindrical or shaped wheels
Drilling	Making of cylindrical forms	Drills
	Passage ways for piping	Core drills
	Anchoring holes	
Cutting at workshop or on site		Sawing machines
		Transportable sawing machines
Drilling on site		Transportable drilling machines

Standard single circular saw blades are very efficient tools, with the material to be sawn determining the choice of bond and diamond type, as well as the sawing parameters. The power of the spindle motor must be appropriate to the saw blade diameter and the forces created in the sawing process. For example, for 500–550 mm diameter saw blades Tyrolit KG [207] recommends sawing machines with 11–15 kW motors. The greater the blade diameter, the more powerful is the machinery required: for 1 m blade diameter it is 37–45 kW, and for 1.5 m diameter — 45–75 kW. At the same time, an ample coolant supply is necessary. The recommended coolant flow rates for the above saw diameters are, respectively, 15–20, 40–50, and 60–70 dm^3/min.

Depending on the stone being cut, the recommended peripheral speeds are: for high-quartz granite — 25–30 m/s; for low-quartz granite — 32–40 m/s; marble — 40–50 m/s, travertine — 50–60 m/s. Very high blade peripheral speeds of over 60 m/s require a different initial tension of the blade blank to attain proper stability.

The cutting speed has a marked effect on blade life and for that reason, the speed recommended by the blade supplier should never be exceeded. Thus, when sawing high quartz granite the recommended cutting speed is 100–600 cm^2/min, while for low quartz granite it is up to 800 cm^2/min [63]. The recommended speed for marble, artificial stones and ceramics is in the range 1500–3000 cm^2/min, and for travertine is 1500–3500 cm^2/min. Excessively fast cutting leads to premature pull-out of individual diamonds from the matrix. If, on the other hand, the cutting speed is too low the diamond tips will be worn smooth which causes an increase in the cutting forces and in the power consumption. Glazed diamond segments must then be sharpened on sandstone or some other abrasive material in such a way that the infeed is low and the feed rate high, i.e. higher matrix wear is induced on purpose to make new diamond points available for use.

Circular blade assemblies are relatively more efficient than single blades. For example, a set of 24 blades makes it possible to slice a 5–6 m^3 block of stone into slabs within one hour. Blade diameter is usually up to 1200 mm.

Diamond blades are highly dependable in such applications as the cutting of glass panels, optical glass (prism making) and armoured glass. Use is made here of continuous rim blades typically 500 mm in diameter. According to Diamant Boart [63], the cutting speeds for window glass are 150–200 cm^2/min, for optical glass — 30–100 cm^2/min, and for armoured glass — about 40 cm^2/min. Standard 400 mm diameter blades are capable of cutting about 10 m^2 of armoured glass per hour. One such blade can cut from 25 to 75 m^2 of optical glass, depending on its composition. The recommended blade peripheral speeds range from 20 to 35 m/s with the highest for window glass cutting.

Circular saw blades are often used to cut highly abrasive ceramics and composite materials with varying properties — laminates or honeycomb structures.

5.1.5.2 I. D. saw blades

Electronic components require high precision sawing and surface machining so that the design dimensions are obtained. A special requirement here is very tight tolerances such as parallelism of the walls, surface flatness, absence of edge chipping, etc.

The tools most widely used for slicing monocrystals, e.g. cylindrical rods known as boules or ingots, are I. D. (or annular) saw blades, which effect fast, efficient and high precision cutting with a very narrow kerf width (relatively low losses of expensive material) and with virtually no propagation of structural defects in the work-piece material. I. D. saw blades were initially used for slicing germanium and silicon boules into wafers. These materials still account for the major usage of I. D. blades, but recently there has been an increase in the slicing of materials such as GaAs, GGG, samarium-cobalt, sapphire and quartz.

I. D. saw blades consist of a very thin circular steel core, which is tensioned at its outside diameter (see Fig. 5.16a). In comparison with conventional saw blades, which are mounted at the centre and have their cutting edge at the outer diameter, I. D. saw blades have their cutting edge at the inner diameter. The stability and high tension of the I. D. blade core enable it to cut thinner slices than are possible with conventional saw blades.

I. D. blade cores are usually fabricated from cold-rolled high strength stainless steel. Blade dimensions are a function of the work-piece diameter. At the completion of the cutting stroke, the distance between the crystal top and the I. D. blade clamping ring should be 10 to 15 mm. This indicates a relationship between work-piece diameter and blade I. D. dimensions. The blade core thickness markedly affects blade stability, which in turn influences blade life and "as cut" wafer quality. Since one cannot increase the core thickness without increasing kerf

Table 5.9 — Materials cut with I. D. (annular) saw blades and recommended sawing parameters [218]

Material	Typical blade r.p.m. (100 mm I. D. blade)	Typical load (g/cm cut length)	Automatic feed rate, approx. (microns/sec)	Surface finish (micro-in. C.L.A.)
Gallium arsenide	140	50	30–37	12
Gallium phosphide	140	50	30–37	12
Indium arsenide	120	30	20–30	12
Indium phosphide	120	30	20–30	12
Lead telluride	120	20	6–13	6
Cadmium telluride	120	20	6–13	12
Lithium niobate	160	50	20–30	8
Proustite	120	20	6–13	16
Tri-glycerine sulphate	120	30	13–20	12
Rutile	160	50	30–37	8
Spinel	160	50	30–37	8
Crown glass	160	50	37–45	8
Quartz	160	50	37–45	8
KDP, ADP	120	30	13–20	12
YAG	160	50	30–37	8
Silicon	160	50	30–37	8
Brass	160	40	20–30	6

losses, it is necessary to produce a blade that strikes a delicate balance between material losses and blade stability [218]. Alignment and clamping bolt holes are located at the outer diameter of the blade and their size and location are standardized. The diamond sizes used in I. D. saw blades depend on the material to be cut. For example, the diamond size recommended by Winter, Germany for silicon slicing is D46 (325/400 US mesh). The sawing of graphite and sapphire requires diamond particles of larger sizes (Table 5.9).

5.1.5.3 Frame saw blades

Large blocks of stone extracted from the quarry are usually sawn into slabs for gravestones or building decoration. The tools most widely used for primary sawing or slabbing are frame saw blades with diamond segments, or large diameter circular saw blades. For frame sawing one needs to mount a pre-determined number of straight blades into a frame. The frame reciprocates, and a feed system is applied either by directing the frame down into the block or by moving the block up to the frame. The sawn slabs will be used predominantly for cladding on the walls of buildings or for tiles [5, 218]. A typical frame saw design is shown in Fig. 5.14. Blade length depends on the machine type. The diamond segment specifications are selected according to the material to be cut, the machine type, its power, rigidity and accuracy [3, 28, 79]. Diamond frame saw blades totalling about 100 as a maximum are mounted horizontally or vertically according to the machine type. They are mainly used for slabbing marble, limestone and sandstone. In exceptional cases, they may be used for cutting soft granite. Well designed frame saw systems offer the advantages of uniform thickness slabs with a good surface finish, resulting in lower total sawing costs despite the

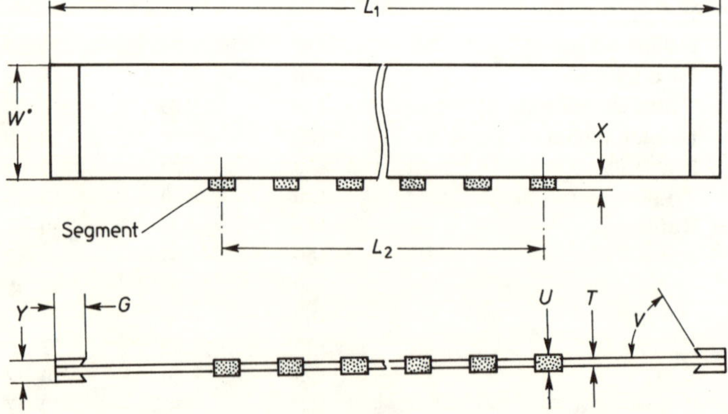

Fig. 5.14 — Schematic diagram of a frame saw design.

comparatively high initial outlay. In typical frame saw blades the segments are mounted along a straight line. Similar in design and operation are frame saw blades where the cutting (armoured) edge is concave, such as the Diamant-Gatterzähne manufactured by Ernst Winter, Germany [218, 219].

5.1.5.4 Band saw blades

Diamond band saw blades are a fairly rare type of tool. They make it possible to cut two oppositely mounted stone blocks at a time. A typical design is shown in Fig. 5.15. Diamond segments are mounted on the edge of a special steel band which is set in circular motion. Machines of this type are used on soft and also on some hard, low-quartz stones [218].

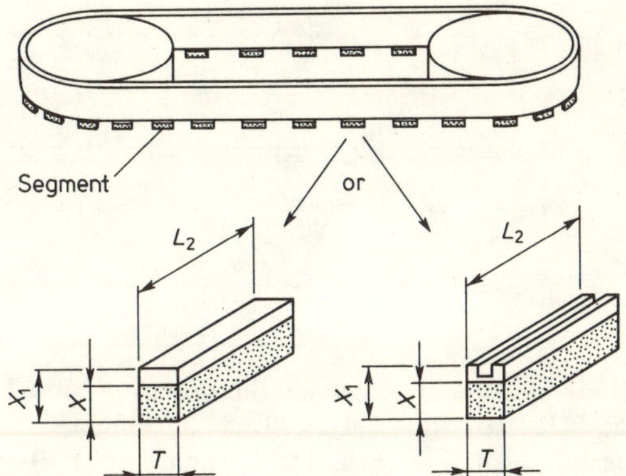

Fig. 5.15 — Schematic diagram of a band saw design [218].

5.1.5.5 Dicing blades

On account of increasing miniaturization and more compact integration of electronic circuits, the wafers cut from monocrystals with I. D. saws

Table 5.10 — Dicing blades and applications [114]

Dimensions of dicing blades				
O.D. (mm)	I.D. (mm)	Thickness (mm)	Diamond size (μm)	Remarks
50 ~ 52	40	0.015 ~ 0.050 ± *	3 ~ 10	Plated bond, for Si, GaP, GaAs
48 ~ 55	25.4 ~ 40	0.10 ~ 0.30 ± **	2 ~ 50	Resin bond (electrically conductive) for Si, ferrite, liquid crystal glass
52 ~ 55	40	0.10 ~ 0.30 ± **	2 ~ 50	Metal bond, for liquid crystal glass

* 0.002 ~ 0.003 mm. ** 0.005 ~ 0.010 mm.

need to be diced into chips. Dicing blades permit kerf widths of as little as 15 μm. Typical dicing blade designs and dimensions are shown in Fig. 5.16, while Table 5.10 gives further practical advice.

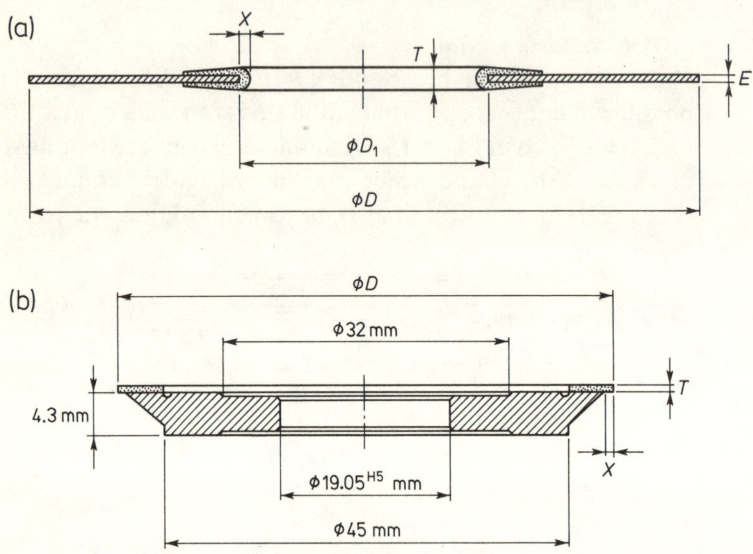

Fig. 5.16 — Typical design and dimensions of: (a) I.D. saw blades; (b) dicing blades [218]. D — from 220 to 690 mm, D_1 — from 83 to 253 mm, E — from 0.1 to 0.2 mm, $T(\pm 0.01)$ — from 0.22 to 0.29 mm, $T(\pm 0.003)$ — from 0.025 to 0.080 mm, X — from 0.1 to 0.2 mm, $X\,(\pm 0.05)$ — from 0.50 to 0.75 mm.

5.1.5.6 Diamond wire

So-called diamond wire comprises a stainless steel multi-strand cable on which the diamond "beads" are placed, together with the spacers and the positioners (Fig. 5.17).

The wire and bead specifications are selected according to the characteristics of the stone or concrete to be cut. Nowadays diamond impregnated beads are used to cut virtually any type of stone. The wire should be used with a machine of 40 h.p. minimum motor power and adequate coolant flow. Diamond electroplated beads are ideally suitable for cutting soft stone (marble) or where the power availability is 25 h.p. or less and the coolant flow is limited. The sawing speed depends mainly on the stone being sawn, e.g. 1–2 m^2/h on granites, 4–7 m^2/h on slates, 3–5 m^2/h on hard marbles, 10–15 m^2/h on travertines, 8–15 m^2/h on crystalline marbles, and 20–30 m^2/h on tuffs.

Diamond wires may be used for:

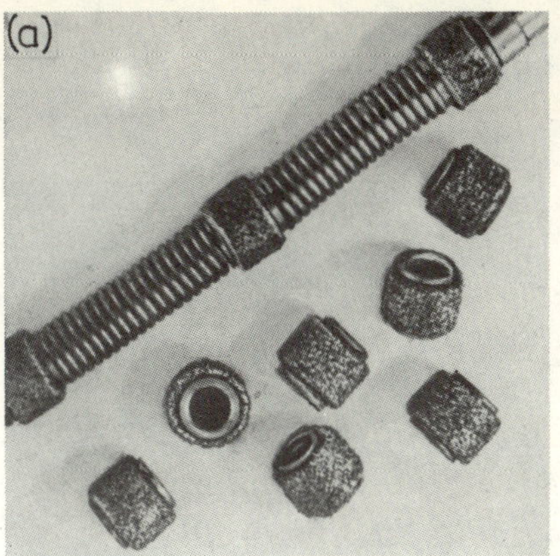

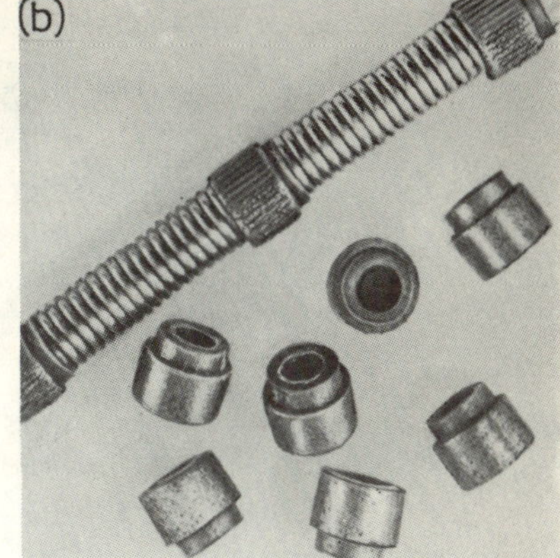

Fig. 5.17 — Diamond wires as manufactured by Diamant Boart. Diamonds bonded by: (a) powder metallurgy; (b) electroplating (courtesy of Diamant Boart SA, Belgium).

— exposing a large quarry face area prior to the extraction of blocks,
— extraction of blocks, by vertical, horizontal or oblique cutting,
— dressing blocks to the desired size, e.g. for frame sawing,
— cutting very large areas, sometimes with varying angles, in reinforced concrete.

The process of extracting blocks from the quarry with diamond wires involves a number of stages. The first consists in drilling vertical and horizontal holes into which to insert the wire. The wire ends are threaded through and linked up then tensioned on the pulley which will later set it in motion. The wire pressure on the stone and, by the same token, the cutting speed are controlled by adjusting the wire tension and its rate of travel. Typical diamond wires are 5, 10, 15 and 20 m long but sections can be joined together to make almost any length. The bead diameter is generally about 10 mm and there are usually some 30–35 beads per metre length of wire. The diamond sizes recommended for beads are 40–60 US mesh [63].

Another related kind of cutting tool is a much finer single wire electroplated with diamond, which may be used for high precision jobs such as cutting out specimens for metallurgical investigations, semi-conductor sawing, etc. [63].

5.1.6 Diamond laps, files and foils

Diamond lapes and files are used for localized finish machining, which may be effected by machine or manually. A lapping tool consists of a substrate impregnated, charged or electroplated with fine diamond particles. Resin and metal bonds are employed (Fig. 5.18). Such tools are

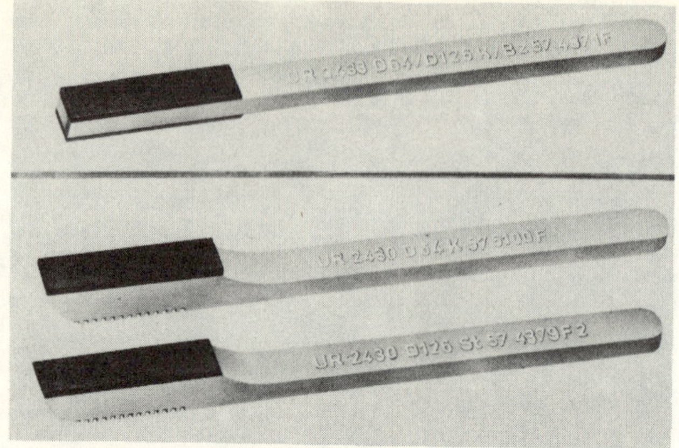

Fig. 5.18 — Examples of diamond lapping tools (courtesy of Joh. Urbanek Co. KG Germany).

particularly useful for removing burrs, bevelling sharp angles, and sharpening of cutting tools.

Diamond files are used for manual finishing operations on blanking dies, die blocks, and mould parts made of sintered tungsten carbide or hardened steels (above 55 HRC). Diamond files come in a wide variety of shapes (Fig. 5.19). Among others there are chasing files, grooving files, etc. However, files should not be used for filing indexable carbide inserts of turning tools [208].

Abrasive products consisting of metal (e.g. stainless steel or non-ferrous metal) foil impregnated about 0.1 mm thick with fine diamond are suitable for manual or machine lapping. This type of product comes

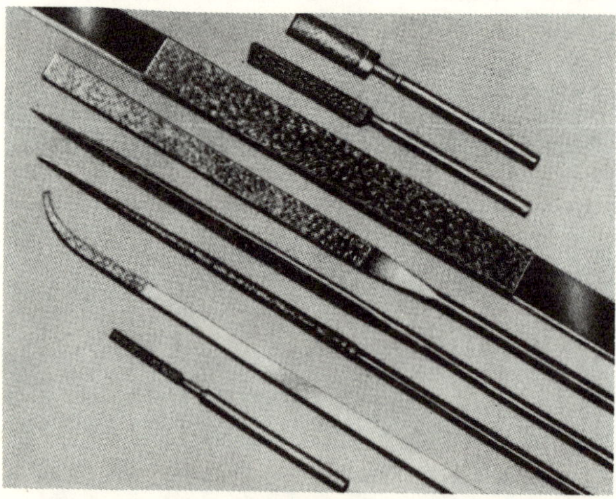

Fig. 5.19 — Examples of diamond files (courtesy of Joh. Urbanek Co. KG Germany).

in sheets, strips or discs, which may be glued onto holders or mounted in clamping rings.

5.2 DIAMOND CLOTHS AND PAPERS

So-called coated abrasives consist of an elastic (e.g. cloth or paper) backing with a layer of abrasive. The production procedure involves spreading an adhesive layer onto a watertight backing such as special grade paper, cloth, fibre, metal foil or fine plastic mesh, in which the abrasive particles are then fixed. Various kinds of resin or other high viscosity adhesives are employed for the purpose, with resin bond products usually being suitable for both dry and wet operations. The usefulness of coated diamond abrasive products is determined by their shape and dimensions, the kind, size and concentration per unit surface area of diamond, the backing type and adhesive employed [4]. Coated diamond abrasive products are manufactured in the form of sheets, belts, endless belts, radial discs and — less commonly — as abrasive flaps for polishing wheels. The diamond quality used for coated products is about the same as for resin bond wheels. Mesh sizes — usually without metal cladding — are typically in the range from 6 μ m up to about 50 US mesh. What is essential is size uniformity, which determines the surface finish attainable. Fine grains hidden by coarse grains remain inactive, while the presence of a single coarse or over-size grain in what is supposed to be a fine grain product may completely disqualify the work for not meeting the surface finish requirements. The extent of particle covering by the glue (the ratio of the glue layer thickness to the particle size) depends on the glue properties and may be varied with the intended application of the product: the value ranges from 0.2 to 0.8 [4].

Diamond abrasive belts may be used for a range of hard surface grinding and polishing operations, including grinding of ceramics platens and carbide die polishing. Diamond sheets are sometimes used for superfinishing chilled iron rolls. These sheets are easily mounted on conventional roll superfinishing equipment. The cloth is held in place by the pressure of the superfinishing head. The back-up can be a piece of pre-shaped wood, a felt, hard or soft rubber pad, or even the normal superfinishing stone after it has been shaped to the roll. Diamond sheets can often get the finishing job done much faster than a conventional abrasive stone, in a predictably timed, reproducible manner.

Cloth backed diamond abrasive discs may be used on surface grinding or polishing machines. The performance of such products is much the same as for metal backed discs. Cloth backed discs are also versatile enough to be used on portable sanders or, in combination with a back-up pad, on rotary hand tools (Fig. 5.20).

Fig. 5.20 — Examples of diamond coated products (courtesy of 3M Company, USA).

Coated diamond abrasive products are also supplied (in some countries) in the form of rolls. Many users convert the rolls to non-standard tool shapes, especially suited to their own operation or to in-house equipment.

5.3 DIAMOND COMPOUNDS, SLURRIES AND AEROSOLS

A common utilization of micron (and sub-micron) diamond powder is in lapping and polishing compounds, suspensions and aerosols. Products based on 60-micron diamond particles are absolutely indispensable today for the preparation of metallographic specimens [9]. They are also standard tools for the preparation of mineralogical specimens, especially thin sections of rocks, ceramics and refractory materials, and they may also be used in the finishing of gemstones, including diamonds [27]. Polishing with diamond compound produces no shape distortion or strains in the sub-surface area, and thus the specimen is fully representative of the bulk material [62].

A diamond compound is prepared by mixing micron diamond with a so-called carrier fluid. As a rule, diamond concentrations range from 0.5 to 20% [9] with the diamond content decreasing with decreasing particle size. A decrease in the diamond concentration results from the increase in the particle surface area with diminishing size. For commercial distribution, compounds are typically packaged in injection syringes or jars containing from 1 to 20 g of compound, while aerosols generally come in cans of 250 g.

The carrier for compounds and suspensions is usually prepared by mixing together a number of substances at up to 100°C, the basic components including resins, fats, alcohols, hydrocarbons, ethers, carboxyl acids, and soaps. To ensure permanent dispersion of the diamond particles, detergents, emulsifiers and preservatives are added. Compounds can be rendered chemically passive or active relative to the work material. Not infrequently compounds contain — in addition to diamond — other fine grain abrasives including boron carbide, silicon carbide, aluminium oxide or other conventional polishing media such as ferric oxide or chrome oxide. Diamond compound manufacturers dye the carrier to denote diamond particle size for easy identification under industrial conditions. Depending on the carrier composition, the compounds present a wide range of properties and consistencies, features which decide the distribution in the course of machining. In terms of consistency, compounds are produced in liquid, semi-liquid and waxy states. Another important practical characteristic of the carrier is the possibility of absorbing the swarf and washing it away from the work surface. From this point of view, diamond compounds are generally classified as water or oil soluble. So-called universal compounds are soluble in water, kerosene, alcohols and workshop oils.

Suspensions and compounds that are unstable in respect of homogeneity are usually prepared by the equipment operator himself immediately before use. One common procedure is to mix dry diamond micron powder with solid or liquid grease or the solvents that are always on hand in the workshop.

The best results in terms of surface finish are obtained if the lapping and/or polishing process takes place in several stages, with increasingly finer diamond particles introduced at each successive stage [9, 118]. Diamond compounds, suspensions or aerosols generally require the use of a lapping plate, disc or cylinder as the work-piece support. In Fig. 5.21 the results of tests using different diamond particle sizes and different laps are presented. Hard materials are best as laps for coarse micron diamond particles; steel or cast iron laps are preferred for rapid stock removal. On the other hand, the best results in terms of surface finish are

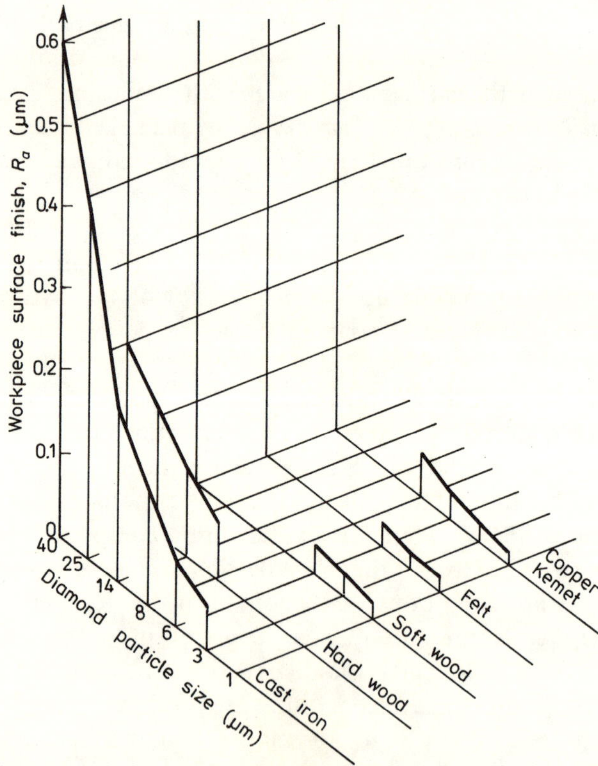

Fig. 5.21 — Effects of working with diamond pastes of different granularity and mounted on different supports [9].

obtained when soft lap materials are used, such as wood, felt, nylon, leather, and non-ferrous metals. With hard lap plates, abrasion takes place predominantly as a result of diamond particle rolling. With soft lap plates, on the other hand, there is little or no particle rolling as the hard particles soon get embedded in the support. Such 'charging' with very fine, ultrahard particles is of greatest advantage when a mirror finish is

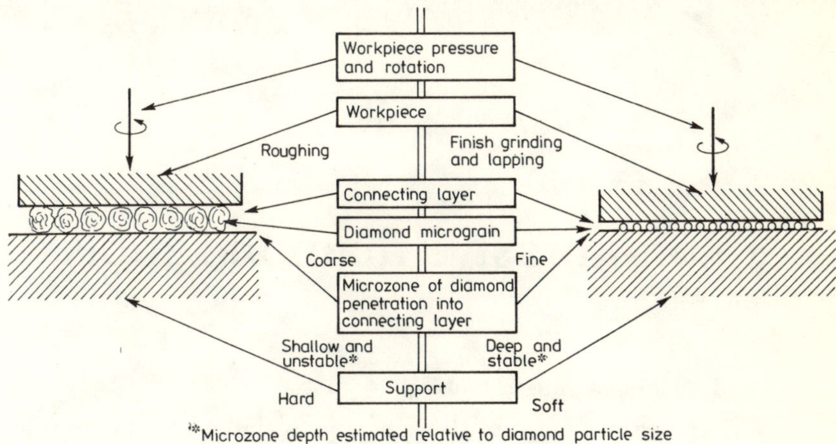

Fig. 5.22 — General recommendation for lap selection in abrasive-polishing operations with diamond pastes [9].

required. However, hard laps are highly wear resistant while soft ones are definitely not. The proverbial 'perfect compromise' may be a combination of two (or more) lap materials in one lap plate (Fig. 5.22). Thus the best lap plate materials for many polishing and superfinishing operations are the so-called Kemets developed by Engis Ltd in the UK, in which softer materials in the form of pellets are combined with a harder material [118].

6

Non-abrasive diamond tools

6.1 Diamond cutting tools

Diamond cutting tools (or single-point lathe tools) comprise a large class of products which, in terms of design and edge shapes, are not too much unlike tungsten carbide, alumina ceramic or high speed steel cutting tools. Depending on the stock removal requirements and work-piece material, either single crystal natural diamond or synthetic polycrystalline diamond inserts may be used. Such cutting tools may be used for turning, boring, milling as well as cutting-off and slitting. Depending on the tool design, there may be a single diamond point as with turning tools, scribers, burnishing tools, etc. or there may be several cutting points, as with milling cutters. The cutting point may be permanently fixed (brazed) in the holder or it may be mechanically clamped.

Products of still another kind which can no longer be regarded as a special type of diamond tool are those with a tungsten carbide substrate covered by a thin diamond layer, as earlier described in Section 2.2.4. Such a thin but hard-wearing diamond layer makes for longer tool life [101, 188].

The shape of a diamond cutting tool depends on a number of factors, chief among them being the machining mode and the machine tool type, the physical and chemical properties of the work material, the method of diamond crystal mounting, and the quantity and size of the diamond(s) employed. A wide variety of special purpose machine tool designs and cutting tool tip shapes are in use. For example, taking into consideration diamond orientation relative to the tool holder axis alone, the following cutting tool types are distinguished: turning, boring and fly-cutting tools, milling cutters, radial tools (concave and convex), etc.

Metalworking drill tools come in the following types:

— drills in which one or several natural diamond crystals or PCD elements are mounted on the end face;
— twist drills in which the active bit is armoured with a PCD insert. Such drills are similar in design to the conventional cemented carbide drill bits. A PCD piece is brazed into the steel drill shank.
— microdrills, in which the active part is entirely shaped of PCD material, e.g. SYNDITE Microdrill Blanks. Such microdrills are used for making tiny holes in thin plates such as printed circuit boards (PCBs) as commonly employed in electronics.

In similar fashion, there is a wide variety of milling cutter and saw blade designs. Table 6.1 lists the tool angles and radii for Soviet natural diamond cutting tools [189]. Because of natural diamond's anisotropy (cleavage planes), it is advisable to use smaller rake and clearance angles,

Table 6.1 — Diamond tool geometry for machining various material [49]

Work-piece material	Tool rake angle, γ	Clearance angle, α	Cutting edge angle, $\varkappa$	Auxiliary cutting edge angle, $\varkappa_1$	Tool corner radius (mm)
Brass, copper, aluminium alloys, plastics	0–−3°	8–12°	30–90°	0–10°	0.2–0.8
Bronze, hard aluminium alloys, titanium	−3–−8°	6–8°	30–90°	0–10°	0.2–0.8

and longer corner radii and angles, compared with carbide tools. Thus when machining hard materials, the angle between the main clearance and auxiliary cutting edges should be no more than 100° [139]. For most turning operations, the rake angle is generally zero, or at most plus 5–10° (for better chip removal and to minimize vibrations when machining e.g. discontinuous surfaces). The clearance angle selected also depends on the work material [218]: with relatively soft (e.g. non-ferrous) work materials it may be as much as 8–12°, while for harder materials (e.g. ceramics), smaller angles are better. In most turning tools, the clearance angles are zero or negative [139, 208, 219]. As a rule, the greater the clearance angle, the lower the cutting forces and, with a shallow depth of cut and modest feed, the main and auxiliary cutting edges hardly become involved in the cutting action. Consequently, both the work surface finish and the cutting forces depend solely on the tip edge radius. When O. D. or I. D. cylindrical surfaces are machined, the cutting edge should coincide with the central part of the diamond [139, 208, 219].

The lathe cutting tool, milling cutter, etc. employed will perform well and wear slowly if the machine tool-cutting tool-work-piece system is sufficiently rigid. The cutting forces with diamond tools are relatively low, which permits the precision machining of low-rigidity materials (e.g. rubber, plastics) and work-pieces (e.g. thin-walled tubes). The surface finish attained depends to a large extent on the machining parameters and on the tool cutting edge geometry (Table 6.2). A high cutting speed may carry the risk of undesirable machine and/or tool vibrations, while a rapid feed rate adversely affects the surface finish. Shallow depths of cut, i.e. of the order of several hundreths of a millimetre, should be used whenever a high quality finish is required. In such cases it is usually advisable to employ e.g. a carbide or ceramic tool for rough machining, leaving a minimal allowance for diamond finishing [208, 219].

When machining with diamond-tipped cutting tools at depths of cut exceeding 0.1–0.15 mm, dimensional accuracy and surface finish are often good enough so that no subsequent lapping and polishing are necessary. If the surface finish starts deteriorating, the usual cause is tool chipping. Natural diamond-tipped tools are sometimes prone to damage on a cutting edge or corner. To minimize this hazard, the tool tip should be relapped sufficiently early, thereby to eliminate the effects of wear.

A cutting tool is used correctly if it is brought into contact with the work and retracted at full work rotation. Natural diamond-tipped cutting tools may be used dry or with a coolant. In the former case, efficient chip disposal (e.g. by air) is essential; the chips can be blown away or sucked by partial vacuum from the cutting area. When a liquid coolant is employed, care should be taken not to direct a stream of cold fluid directly onto the hot tool tip: a mist system is preferable.

PCD is becoming increasingly popular as a substitute for natural diamond. The main advantage of synthetic PCD is the isotropy of its physical properties. Moreover, literally any shape can be cut by wire EDM or by laser from as-synthesized PCD discs. The polycrystalline diamond layer on carbide-backed inserts resists chipping and wear. The cemented carbide substrate gives the insert toughness and mechanical shock resistance. Because of their high resistance to wear, shock and vibration, PCD cutting tools are being successfully used in tough machining applications, e.g. with interrupted cuts. Materials most efficiently machined with SYNDITE turning tools and milling cutters

Table 6.2 — The machining parameters and the tool cutting edge geometry

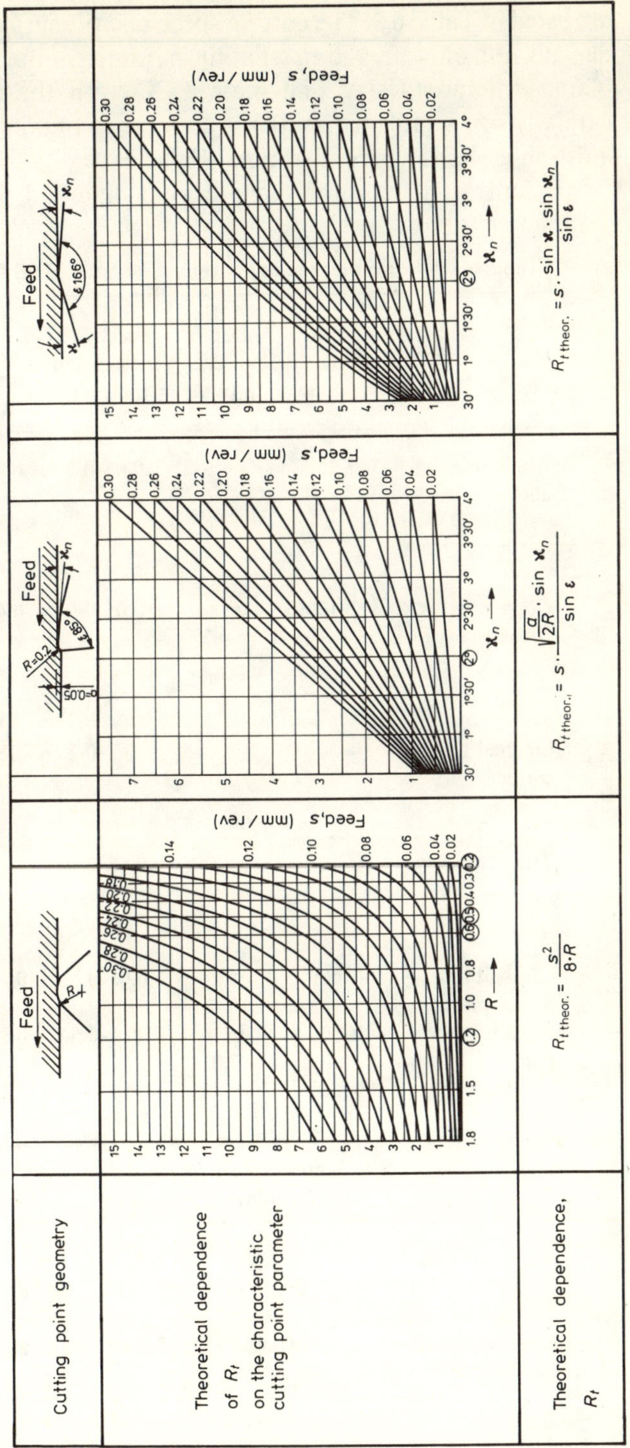

Cutting point geometry	Feed, R	Feed, R=0.2, a=0.05, ε 85°	Feed, ε 166°
Theoretical dependence of R_t on the characteristic cutting point parameter	R → (1.8, 1.5, 1.2, 1.0, 0.8, 0.6, 0.5, 0.4, 0.3, 0.2); Feed, s (mm / rev)	$\varkappa_n$ → (30′, 1°, 1°30′, 2°, 2°30′, 3°, 3°30′, 4°); Feed, s (mm / rev)	$\varkappa_n$ → (30′, 1°, 1°30′, 2°, 2°30′, 3°, 3°30′, 4°); Feed, s (mm / rev)
Theoretical dependence, R_t	$R_{t\,\text{theor.}} = \frac{s^2}{8\cdot R}$	$R_{t\,\text{theor.}} = s\cdot\frac{\sqrt{\frac{a}{2R}}\cdot\sin\varkappa_n}{\sin\varepsilon}$	$R_{t\,\text{theor.}} = s\cdot\frac{\sin\varkappa\cdot\sin\varkappa_n}{\sin\varepsilon}$

are listed in Table 6.3. The cutting speed and the surface finish attainable depend not only on the machining parameters but also on the PCD blank structure. Using fine grain PCD with the diamond particles carefully size segregated makes it possible to obtain a superior surface finish in a shorter period of time (Fig. 6.1).

In contrast to natural single crystal diamond, PCD inserts wear

Table 6.3 — Recommended applications of SYNDITE PCD tools [55]

	Work-piece material	Cutting speed, v (m/min)	Depth of cut, a (mm)	Feed rate, s (mm/rev) s_t (mm/tooth)	Tool corner radius (mm)	Remarks
Turning	Aluminium alloys	300–1000	up to 10.0	0.05–0.5	0.4–1.2	Depends upon machine rigidity and available spindle power
Turning	Copper, brass, bronze alloys	300–1000	up to 10.0	0.05–0.5	0.4–1.2	Depends upon machine rigidity and available spindle power
Turning	Sintered tungsten carbide	10–30	up to 2.0	0.1–0.2	4.76–6.35	9.5 or 12.7 mm diameter inserts—rigid machine
Turning	Titanium alloys	50–100	up to 2.0	0.05–0.1	0.8–1.2	Not all alloys can be machined successfully
Turning	Glass-fibre reinforced plastics	100–600	up to 5.0	0.05–0.5	0.8–1.2	
Turning	Silica-filled plastics	400–800	up to 2.0	0.1–0.5	0.8–1.2	
Turning	Ceramics	100–600	up to 2.0	up to 2.0	0.4–1.2	
		Cutting speed (m/min)	Feed rate (mm/rev)	Depth of cut (mm)		
Milling	Aluminium alloys	500–3000	0.1–0.5*	up to 5.0		
Milling	Copper, brass and their alloys	200–1000	0.1–0.5*	up to 2.0		
Milling	Chipboard and fibreboard	2000–3000	1.5–2.0*	up to 15.0		

* mm/tooth.

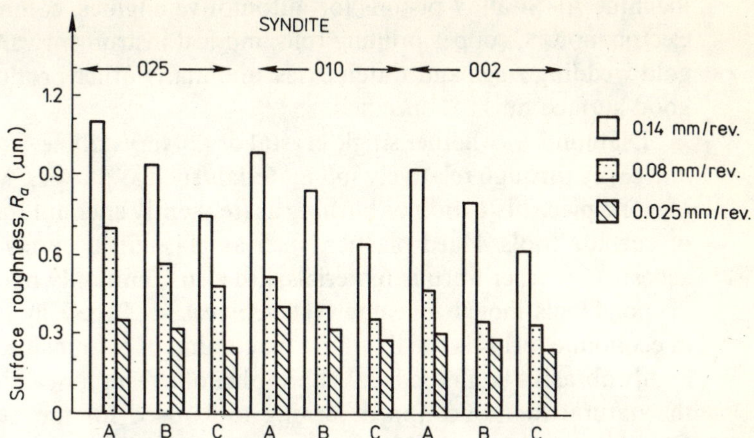

Fig. 6.1 — Dependence of work surface roughness and turning speed on the PCD grade [1]: A — rough ground, B — finish ground, C — finish ground and lapped.

mainly by abrasion. Obviously enough, an excessively worn tool will result in inferior performance and for that reason, natural diamond and PCD tools should be relapped as soon as deterioration of the cutting edge sharpness is observed. The isotropic structure of synthetic PCD makes it possible to obtain tool shapes unthinkable in the case of natural crystals (Fig. 6.2).

Diamond cutting tools are used [55] on high and ultra-high precision lathes and boring machines for the machining of a variety of non-ferrous metals and alloys, including copper, brass, bronze, silver, gold, platinum, the alloys of aluminium, magnesium, titanium and zinc. As already noted, natural diamond turning and boring tools can produce a very high quality surface finish on these materials which needs no further lapping or polishing. Where applicable, a work-piece can be rough-turned with a carbide tool and finish machined on the same lathe

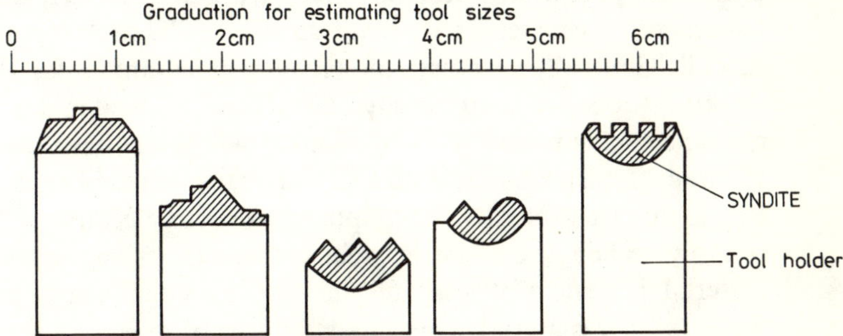

Fig. 6.2 — Possible shapes of turning tools with SYNDITE PCD inserts [55].

with a natural diamond or PCD tool. PCD tools are nowadays used to machine Al–Si alloy pistons for automotive engines, commutators for electric motors, copper printing rolls, musical instrument valve cylinders, gold wedding rings and watch cases, and many other products where a good surface finish is specified.

Diamond — whether single crystal or polycrystalline — cuts cleanly and easily through relatively soft materials such as plastics, with no signs of work-piece distortion or burning as frequently encountered with HSS or carbide tools. Hard plastics, such as glass filled epoxy resins, and asbestos or other fibrous materials, are also commonly machined with diamond tools, though for entirely different reasons. Especially advantageous in economic terms is the diamond machining of composites such as the highly abrasive laminated glass fibre/phenolic resin range. For example, the natural diamond-tipped turning tools used for the machining of Gemini rocket heatshields cut some 56 linear kilometres before they needed resharpening. By contrast, sintered carbide tools had to be resharpened before they had cut 1 linear kilometre.

The introduction of PCD tools has made a major impact on the furniture industry, especially in automated operations involving long production runs and, in particular, when laminated, plastic coated boards are being machined. SYNDITE or COMPAX PCD inserts increase tool performance by 100 times and more as compared with similar design tungsten carbide tools. Typical PCD tools for woodworking include saws, milling cutters and routers with a variety of sometimes quite complex shapes. These are especially useful for the single pass edge profiling of chipboard and fibreboard [50]. PCD tools for the furniture and wood-processing industries are nowadays manufactured by over 20 companies, among others Lach-Spezial-Werkzeuge GmbH of Germany and the USA. Lach-Spezial PCD cutter blocks only require resharpening after cutting between 100 and 150 linear km of medium density fibreboard (MDF), and sometimes they stay sharp for as much as 200 linear km [117]. Records, however, are being broken all the time. Different tool geometries are required when PCD tools replace tungsten carbide. Generally speaking, the recommended angles for PCD woodworking tools are: cutting edge back rake 5°, wedge angle 75°, and clearance angle minus 10°, while for carbide tools these angles are 15°, 55°, and 20° respectively. With PCD when the included angle is less than 65°, there is a tendency for chipping to occur because of the lack of support. When the angle is greater than 75°, shearing of the work material is generally less and the cutting edge has a tendency to 'bulldoze' the material, causing edge finish problems [49].

An entirely different example of the use of PCD tools is in the mining industry. Apart from full-hole and core drill bits, the working parts of some tunneling machines may be armoured with PCD inserts, e.g. of the SYNDRILL or STRATAPAX type, which substitute for the standard sintered carbide parts. Because of the inhomogeneity of most rock formations, the PCD elements not only cut but also shear the rock in compression which is a more efficient process. Such equipment is employed, among others, in coal mining on a semi-experimental basis [49]. De Beers has developed a special PCD tool element for this type of operation, called SYNPICK.

6.2 SINGLE CRYSTAL DIAMOND CUTTING, SCRIBING AND BURNISHING TOOLS

At one time a very common type of diamond-tipped tool was the glass 'cutter'. They were employed to make a scratch on the surface of a glass

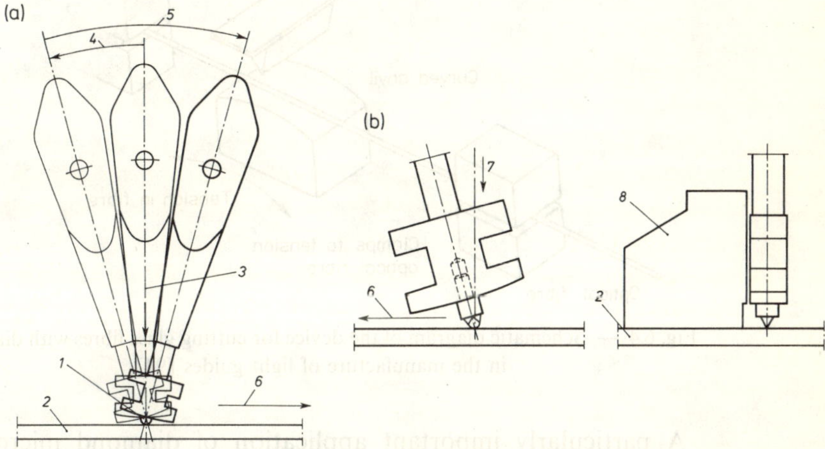

Fig. 6.3 — Scratching a glass plate with a diamond glass cutter for the purpose of separating it [139]: *1* — diamond crystal, *2* — glass plate, *3* — direction of knife application to the plate, *4* — making an indentation in the plate with the knife at about 15°, *5* — bringing the tool back to index along the radius position — with the diamond remaining motionless — and positioning the diamond at about 15° in the direction of scratching, *6* — scratching the plate with the diamond, 7 — direction or depth of scratching, *8* — straight-edge rule.

plate and thus to generate tensile stresses on either side of the scratch. On being appropriately pressed, the glass cracks precisely along the scratch. The tool consists of a handle, a holder, e.g. made of steel, into which a diamond chip is brazed. The way to use such a tool is illustrated in

Fig. 6.3. Only one scratch is allowed [139]. Nowadays, however, the more robust carbide wheel type of tool is mainly used.

Diamond knives are used for cutting biological tissue, in eye surgery, etc. [72]. They are indispensable for the accurate jointing of fibre optics cables — they make it possible to obtain flat cleavage surfaces without any edge spalls. The cleaving operation is conducted on specially designed anvils, as depicted in Fig. 6.4. The scoring point of the diamond tip has the shape of a prism with an apex angle of 30–45°, and the cutting edge is up to 3 mm long.

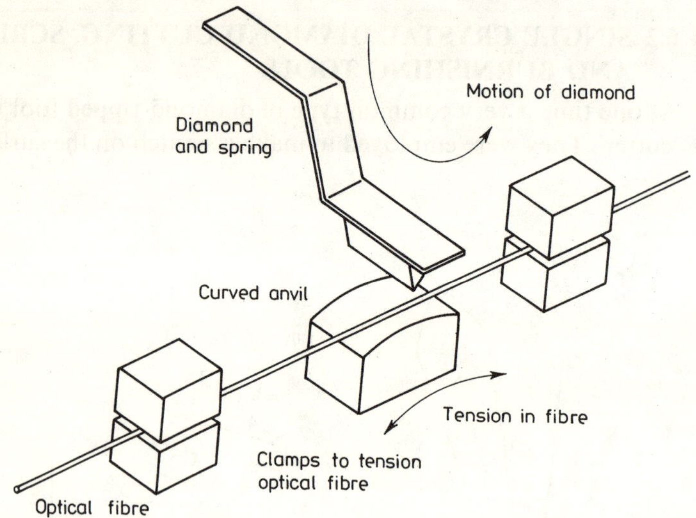

Fig. 6.4 — Schematic diagram of the device for cutting glass fibres with diamond in the manufacture of light guides [94].

A particularly important application of diamond microsurgical knives is in eye surgery and in microtomes for obtaining biological specimens, when living tissue is being cut. Ultraprecision diamond knives make it possible to obtain specimens about 100 Å thick [72]. Knives for microsurgical applications comprise thin diamond plates, e.g. 0.1–0.3 mm thick, 1.0–1.6 mm wide and 100–150 mm long. The blade is sharpened on one side (Fig. 6.5). One company involved on an international scale in the marketing of surgical knives is Microsurgical Administrative Services Ltd. in the UK. Apart from ophthalmic surgery, in particular cataract removal and myopia correction, diamond knives are also being tested in coronary and deep brain surgery. An invaluable development was providing such tools with a source of laser light which facilitates greater precision of the operation. According to

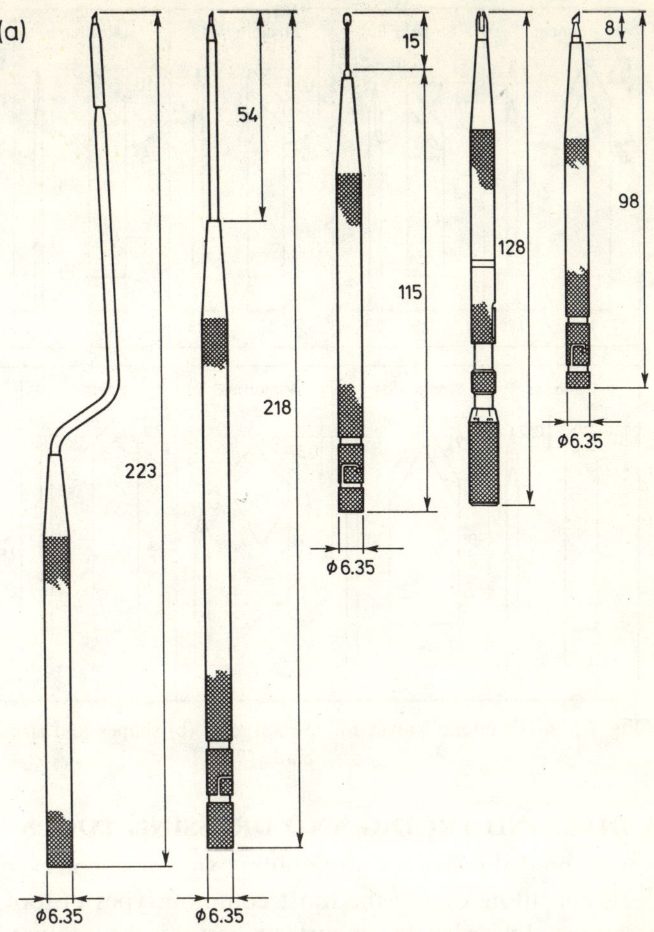

Fig. 6.5 — Diamond knives for eye surgery: (a) with the diamond blade mounted in a holder;

Hoskin [80], a simple combined tool for incision and cautery is one obvious application, but laser-incorporated knives could also have considerable potential in surgery involving, say, cancerous tissue, where the knife would be able to excise the tissue in a dry state to help prevent the cancerous cells from distributing themselves around the body. Another reported potential opportunity is the development of a diamond-tipped laser instrument for use in "welding" blood vessels together [72, 181].

Natural diamond-tipped engraving tools are widely used by professionals and amateurs for the decoration of glass drinking vessels, fruit bowls, paperweights, presentation pieces, etc.

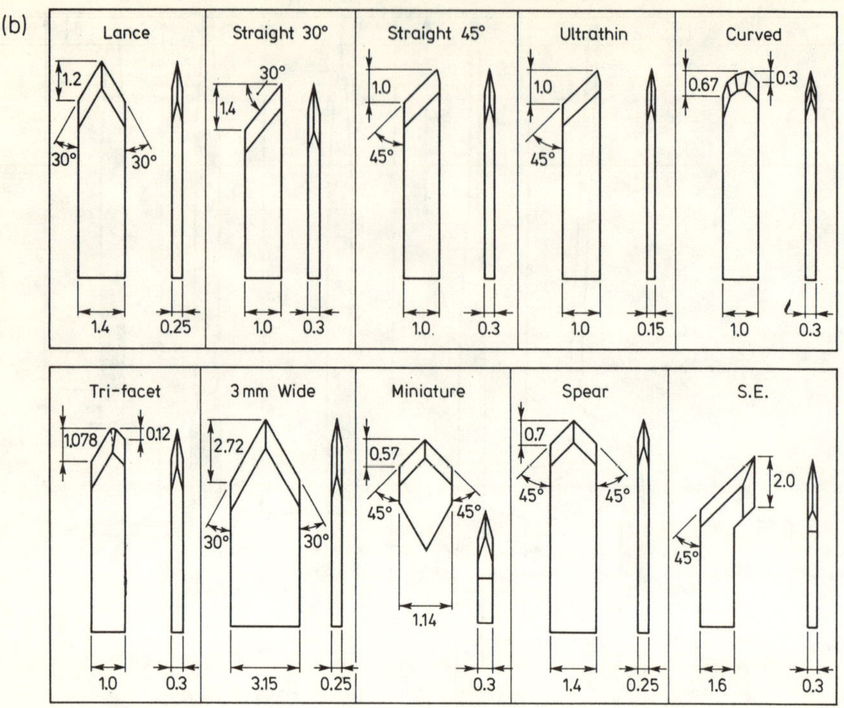

Fig. 6.5 — Diamond knives for eye surgery: (b) shapes and sizes of diamond blade [72].

6.3 DIAMOND TRUING AND DRESSING TOOLS

Conventional abrasive (i.e. aluminium oxide or silicon carbide) grinding wheels constitute one of the most common types of tool for grinding operations. In order to impart or restore the required shape and performance characteristics to such wheels, they are from time to time subjected to two quite separate operations known as truing and dressing. The term "diamond truing" is sometimes used to cover not only the generation or restoration of the geometric form, but also de-clogging (dressing) of the working surface. For obvious reasons, the diamond tool is almost invariably expected to carry out both functions simultaneously. Both operations can be performed manually but preferably by controlled mechanical means. Theoretically speaking, full restoration of the original wheel shape may be attained if a layer as thin as 0.08 mm is removed. In practice, a somewhat thicker layer is removed, even though this does not improve the wheel efficiency: what it does is increase the wheel and tool wear. The amount removed is greater when the truing and dressing operations are performed manually, when large diameter, poorly balanced and/or low hardness wheels are involved, when the machine tool or dressing device components do not have the required rigidity, etc. [71, 96].

Diamond dresser life is defined in terms of the number of truing and/or dressing operations it is capable of performing. With diamond it may be more than 100,000 operations, which more than offsets the relatively high tool cost.

The end result of dressing is the sum of a great number of factors. Because of that, the things to keep in mind when selecting a truing and dressing tool are:

— the abrasive wheel characteristics, i.e. hardness, abrasive type and grit size, etc.;
— the area and geometry of the surface to be dressed;
— the amount of material removed in truing/dressing;
— conditions of the grinding machine and dressing device;
— the operational wheel speed;
— the quantity of parts to be ground and the tolerance requirements;
— labour (where applicable) and tool costs.

On account of the great diversity in shapes and sizes of components generated or finished by grinding, and the widely differing surface finish requirements, a broad variety of dresser designs is available. Practical tool use is facilitated by the national Standards which describe them, such as American National Standard ANS B67-1-1975 and British Standard BS 2002:1973. Generally speaking, the most popular dresser types are:

— the single point dresser,
— chisel-type dressers,
— multi-point dressers,
— diamond impregnated dressers,
— block and rotary truers (dressers).

Single-point diamond dressers for manual use employ selected shaped diamonds, mounted in a metal matrix. Such tools are used to dress and to impart or restore the required geometric shape to general-purpose electrocorundum (aluminium oxide) and silicon carbide abrasive wheels. With dressers of this simple type, the wheel diameter and width, on the one hand, and the diamond crystal size on the other, have to be matched, as shown in Fig. 6.6. Diamond crystal wear should not exceed 1/3 of the initial height — any use beyond that limit may make it impossible to re-shape the crystal for subsequent use. Diamond wear also depends on the crystal orientation relative to the work. The recommended crystal corner positioning is 1–3 mm below the wheel axis and at an angle of 3–15°, while the recommended angle of dresser inclination relative to the longitudinal feed direction is 0–20°, which

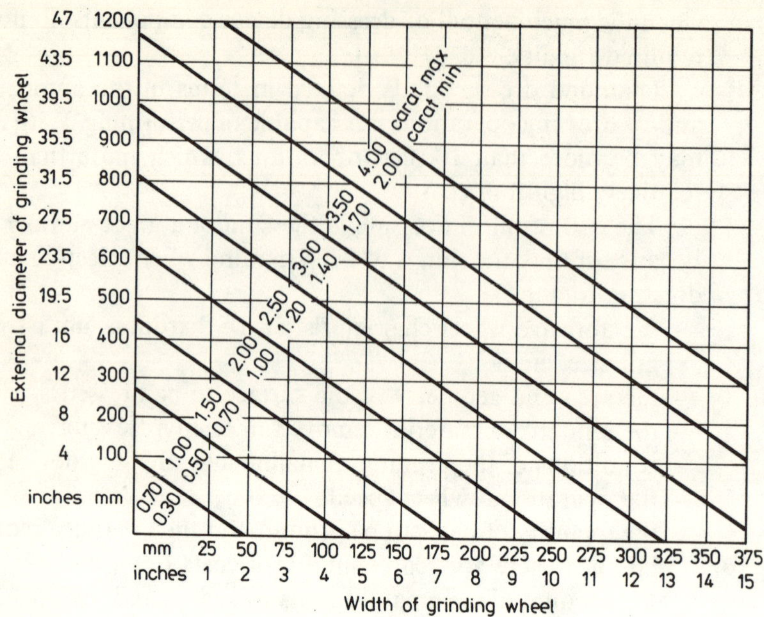

Fig. 6.6 — Wheel width and diameter vs. diamond size for a monocrystal dresser [218].

gives an angle of attack inclination of 0–5° [96]. Recommendations for the use of natural diamond dressers manufactured by PONAR-ŁÓDŹ in Poland are presented in Table 6.4. Tool life is prolonged if the truing/dressing operation is carried out in two stages, i.e. roughing and finishing.

In recent years, it has become increasingly more common to replace natural diamonds with PCD elements, such as e.g. FORMSET or COMPAX [49, 85]. The main advantage of PCD insert wheel dressers is the isotropy of the PCD, which makes repetitive work much simpler for the machine operator. Such tools are used especially for dressing 1A1 shape wheels.

A characteristic feature of single-point diamond dressers is that the tip of the diamond is specially shaped. A shaped edge makes for more precise truing and profiling, and is thus especially desirable where complex wheel shapes and/or high dimensional tolerances are required, e.g. on multi-ribbed thread grinding wheels. The most popular edge shapes are cones, pyramids, and prisms. Depending on the diamond and holder shapes, several types of such dressers are distinguished, commonly referred to by the name of the manufacturer who first introduced them onto the market, e.g. Schaudt, Reishauer, Fortuna. The highest quality diamond dresser stones are used for making polished diamond dressers.

Table 6.4 — Recommended application parameters for single crystal diamond dressers manufactured by PONAR-ŁÓDŹ (Poland)

Technological process				Conditions of diamond dresser use				
Kind of grinding operation	Surface finish		Longitudinal feed, P (mm/min)	Depth of dressing (infeed) (g)				Number of dresser passes without infeed
	Steel			Rough dressing		Finish dressing		
	Heat treated	Mild		mm/passes	number	mm/passes	number	
Centre-type grinding with longitudinal feed	1.25	—	0.3–0.4			0.01		—
	0.63	1.25	0.2–0.3	0.02–0.03	2–3	0.01	1–2	1
	0.32	0.63	0.1–0.2			0.01		1–2
	0.16	0.32	0.05–0.1			0.005		1–2
Centre-type grinding with cross feed	1.25	—	0.15–0.25			0.1		1
	0.63	1.25	0.08–0.15	0.02–0.03	2–3	0.1	1–2	1–2
	0.32	0.63	0.05–0.08			0.005		1–2
Centreless grinding with longitudinal feed	0.63	—	0.1–0.15			0.1		1
	0.32	1.25	0.08–0.10	0.02–0.03	2–3	0.1	1–2	1–2
	0.16	0.63	0.05–0.08			0.1		1–2

Table 6.4 (cont.)

Centreless grinding with cross-feed	0.63	1.25	0.07–0.15			0.1		
	0.32	0.60	0.05–0.09	0.02–0.03	2–3	0.05	1–2	1–2
Chuck-free grinding with longitudinal feed	—	—	0.10–0.15	0.02–0.03	1–2	—	—	—
ID grinding with longitudinal feed	1.25	2.5	2.0–3.0			0.01		1–2
	0.63	1.25	1.0–2.0	0.02–0.03	2–4	0.005	1–2	2–3
	0.32	0.63	0.5–1.0			0.005		2–3
Surface grinding	1.25	2.5	0.6–0.8			0.01		1
	0.63	1.25	0.4–0.6	0.02–0.03	2–3	0.01	1–2	1–2
	0.32	0.63	0.2–0.4			0.005		2–3
Thread grinding	1.25	—	0.08–0.15	0.01–0.02		0.01		1
	0.63	—	0.05–0.08			0.005		2–3
Profile grinding with automatic control	1.25–0.63	2.5–1.25	0.2–0.4	0.03–0.05	3–5	0.01	1–2	1–2
	0.63–0.32	1.25–0.63	0.1–0.3	0.02–0.04	3–4	0.005	2–3	2–3

The relatively high precision nature of the tool (notably of its diamond point) requires particular care in handling. Such tools are often used on ancillary grinding machine equipment, ranging from simple radius dressing attachments to more complex pantograph (e.g. Diaform, Truepath, etc.) and optical profile equipment (e.g. Optidress). An advantage of polished single point diamond dressers is the small tool/wheel contact area and, consequently, the generation of relatively low cutting forces which results in higher precision in dressing and slower diamond wear. The basic principles of polished single point dresser use are similar to those for diamond dressers set with a rough diamond. The recommended directions of dresser movement relative to a complex shape wheel face are illustrated in Fig. 6.7.

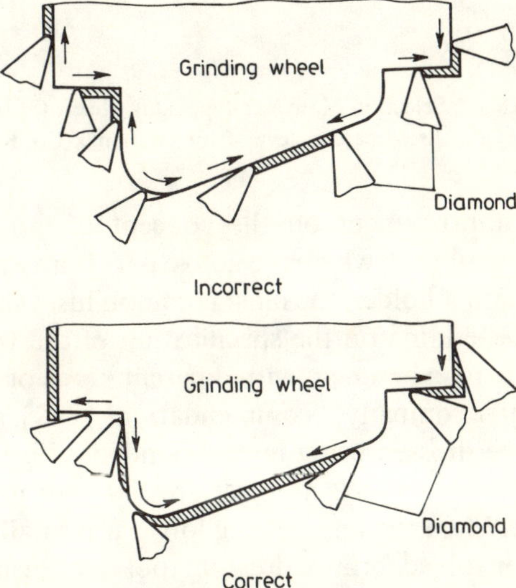

Fig. 6.7 — Recommended directions of motion made by single-point dresser relative to complex profile wheel [71] (similar principles hold in the case of dressing with rib dressers).

For some time there has been a tendency to replace dressers set with large natural diamond crystals with multi-point and diamond impregnated tools. Crystals used for some multi-point dresser designs range from 0.3 to 0.02 ct. The diamonds may be arranged in various patterns or even by layers. They may also be arranged on a disc periphery in one or several rows. The various designs are typically known as cluster type, blade type, rotary type, layer type, etc. (Fig. 6.8). Rotary dressers are especially suitable for dressing wheels used for machining elements of roller bearings, while the layer type is suitable for dressing straight (1A1) grinding wheels.

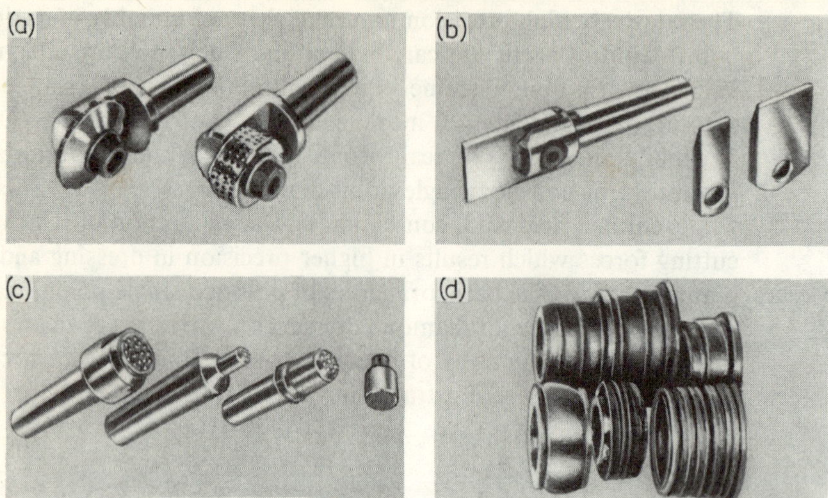

Fig. 6.8 — Selection of dresser types: (a) wheel; (b) blade; (c) cluster; (d) roller dressers (courtesy of Joh. Urbanek Co. KG Germany).

An improvement on the concept of multi-point dressers is the impregnated tool, where the active part is an abrasive inset mounted or brazed into a holder, the most common inset shapes being a cylinder or plate. Depending on the specification of the wheel being dressed, it is advisable to experiment with different kinds of tool. Thus, according to the Winter company recommendations [218], aluminium oxide wheels should be dressed using inserts bonded with tungsten powders, while silicon carbide wheels are best dressed with inserts based on tungsten carbide. When the unit cutting forces are small and when fine abrasive grain is involved, bronze dressing tool bonds can also be employed. The inserts should be used until they are fully worn. Plate-type dressers are suitable above all for dressing wheels for grinding complex profiles such as threads, while dressing tools of cylindrical or prism shape are best for flat surfaces and straight profiles (Fig. 6.9).

Roller and block-type dressers (Fig. 6.8d) are used for truing profiled wheels intended for plunge or creep feed grinding on surface grinders. The dressing block and roller shapes correspond exactly to the wheel contour. Using such tools it is possible to true a profiled wheel or even a set of wheels at one go without interrupting the grinding operation. Roller and block dressers are, however, very expensive and they make economic sense only if used in large batch, fully automated operations. They are particularly suitable for profile grinding operations in the aircraft, automotive, bearing and gear industries. The accuracy requirements for roller and block dressers are outlined in Fig. 6.10.

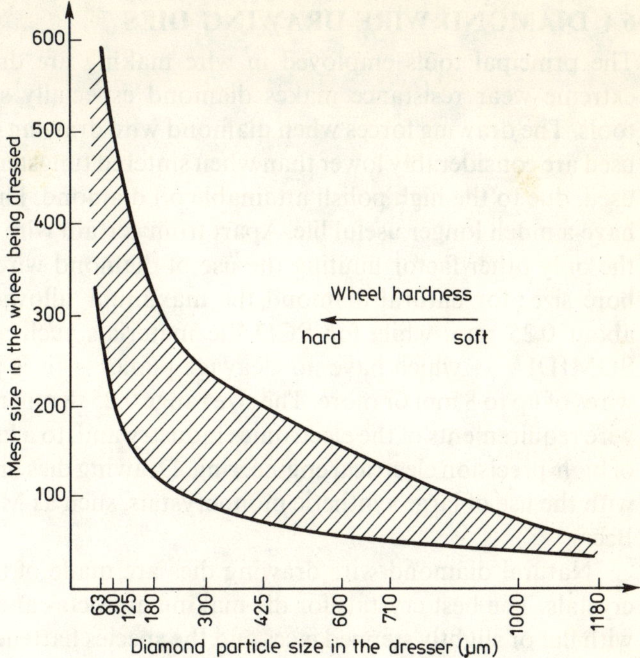

Fig. 6.9 — Diamond particle size in a multigrain dresser vs. wheel hardness, and the abrasive grain size in the wheel [218].

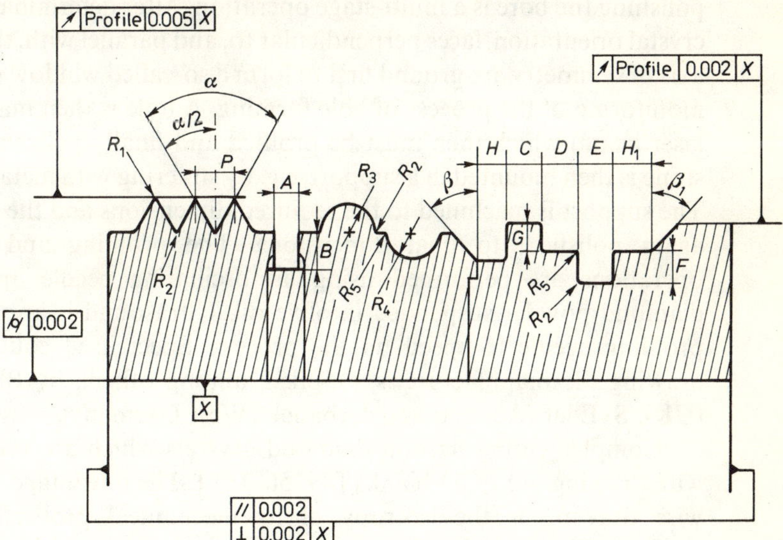

Fig. 6.10 — Typical accuracies attainable for various elements of roller and block dressers [7, 214, 218]: $A = \pm 0.002$, $D = \pm 0.005$, $G = \pm 0.002$, $P = \pm 0.003$, $\beta = \pm 5'$, $R_2 = 0.05$, $R_5 = 0.25$, $B = \pm 0.001$, $E = \pm 0.01$, $H = \pm 0.002$, $\alpha = \pm 5'$, $\beta_1 = \pm 10'$, $R_3 = \pm 0.005$, $C = \pm 0.004$, $F = \pm 0.005$, $H_1 = \pm 0.01$, $\alpha/2 = \pm 2' 30''$, $R_1 = 0.1$, $R_4 = \pm 0.005$.

6.4 DIAMOND WIRE DRAWING DIES

The principal tools employed in wire making are drawing dies, and extreme wear resistance makes diamond especially suitable for such tools. The drawing forces when diamond wire drawing dies (WDDs) are used are considerably lower than when sintered tungsten carbide dies are used, due to the high polish attainable on diamond. Diamond dies also have a much longer useful life. Apart from certain wire materials, about the only other factor limiting the use of diamond wire drawing dies is bore size: for natural diamond the maximum 'allowable' bore size is about 0.25 mm, while for PCD die materials such as SYNDITE or SUMIDIA — which have no cleavage planes — it is possible to draw wires of up to 8 mm or more. The size range 0.25–8 mm meets most of the wire requirements of the electronics industry and, to a large extent, those of high-precision electrical engineering. Drawing dies are now also made with the use of large synthetic monocrystals, such as MONODIE of De Beers.

Natural diamond wire drawing dies are made of near gem quality crystals. The best crystals for die making are octa- and dodecahedrons with flat or slightly stepped faces and the species flattened along the diad and quaternary axes. These axes present the optimal direction of the wire drawing bore in terms of die life. Piercing the diamond and polishing the bore is a multi-stage operation. After determination of the crystal orientation, faces perpendicular to, and parallel with, the selected axis of symmetry are ground first to form a so-called window permitting monitoring of the process of hole forming. A hole is then made (e.g. by laser beam) which then must be profiled and finally polished. The die stone is then mounted in a support, e.g. by sintering with metal powders. The support is machined to the required dimensions and the die bore is again polished, to final dimensions. The profiling and polishing operations are performed using an oscillating needle and micron diamond powder suspended in olive oil. Ultrasound is applied to the profiling needle to speed up the process. Specialized equipment for drawing die manufacture is supplied, among others, by BDWD Co. (UK), S. Eder (Austria) and Urbanek (West Germany).

Complementing natural diamond crystals which are used for fine wire drawing are PCD blanks [49, 56, 161]. The advantage of PCD in wire drawing is the isotropy of its mechanical properties, which considerably simplifies the die production process. Holes in the PCD blanks are made by pulse laser or electric discharge methods [161], the latter method being especially useful in hole shaping. After electric discharge piercing with tube electrodes, solid shaped electrodes are used to complete the die profile. The die geometry is accurate and an acceptable surface texture is achieved. Using appropriately shaped electrodes one may obtain holes with e.g. non-circular (polyhedral) sections, elongated in one or several directions, square or rectangular sections, etc. The polishing time for PCD materials is longer than for natural diamond. Even when PCD surfaces are highly polished, they retain a slightly textured appearance when viewed under high magnification due to their inherent polycrystalline microstructure.

Calibration of the die bore can be carried out in two ways: by high speed mechanical polishing or by oscillating wire [38, 76]. Both methods 'pull' diamond slurry through the bearing area, an action which also blends in the reduction zone. Diamond micron sizes used for WDD polishing are fine, starting at about 3 μm and are applied by experienced technicians who carefully monitor the operation.

A cross-section through a natural diamond wire drawing die is depicted in Fig. 6.11. The exact die profile depends on the drawing

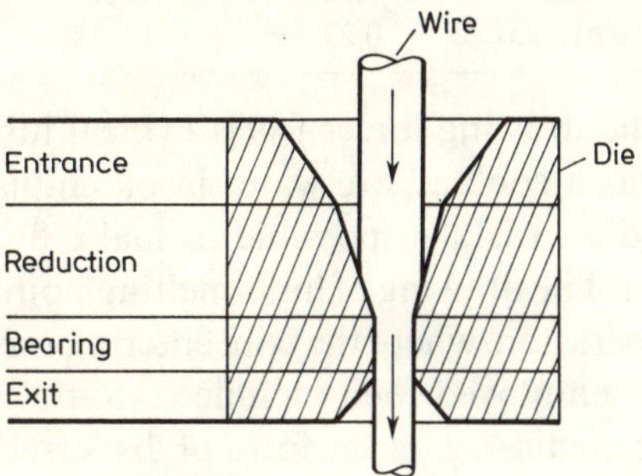

Fig. 6.11 — Names of the different parts of the channel in diamond wire drawing dies.

parameters as well as the wire material. The recommended dimensions of holes for drawing wire from different materials are listed in Table 6.5. Experience has shown that in terms of output capacity, PCD dies are superior to all other kinds. For non-ferrous metals, SYNDIE or COMPAX PCD dies make it possible to draw up to several hundred times more wire than sintered carbide dies. In the larger hole sizes, PCD dies are also several times more productive even than natural diamond dies.

Wire is always reduced in diameter in a number of stages, drawing it several times through increasingly smaller holes to avoid excessive compressive forces which would burst the dies. As a rule the drawing of non-ferrous metal and steel wires is a cold operation, and the friction of the wire against the diamond is the main factor determining die life. The friction can be reduced with suitable lubricants. These soften the outer layer of the metal, which lowers its flow limit and, by the same token,

Table 6.5 — COMPAX diamond blank die geometries for various metals [49]

Application	Die bore sizes (mm)	Reduction angle	Percent bearing length	Percent area reduction	Drawing speed (m/min)
Aluminium	0.16–7.60	16–22°	10–30	18	1200
Copper	0.05–7.60	16–22°	10–30	21	1800
Tin plated copper	0.50–1.80	16–18°	30–40	21	1200
Aluminium/magnesium alloy (5056)	2.05–4.76	16–18°	20–40	22	620
Nickel 200	0.33–1.45	16–18°	30–50	25	300
Tungsten	0.12–0.62	12–14°	40–60	20	60
Molybdenum	0.12–1.02	12–14°	30–50	21	70
Galvanized high carbon steel	0.24–1.05	10–12°	20–40	17	760
Brass plated steel (tyre cord)	0.17–0.96	10–12°	20–40	18	760
Low carbon steel	0.88–2.10	12–14°	30–40	18	760
Ni–Cr–Fe (60:15:25)	0.23–0.91	12–14°	30–40	23	180

reduces the drawing forces [49]. Typical lubricants which, at the same time, act as a coolant, are water-in-oil emulsions. Using diamond wire drawing dies it is also possible to make fine tubes, e.g. from stainless steels [50]. The drawing of high-melting point metals, on the other hand, is a hot operation, where the wire entering the die bore is pre-heated. The lubricants employed then include colloidal graphite and other similar inorganic media, e.g. in the form of dispersed suspensions. The speed of wire drawing is slower for tungsten and molybdenum than for more ductile metals, e.g. copper, or alloys. Heating the wire has an adverse effect on die life. Fine grain PCD blanks, which are thermally stable to 1450 K in an inert or reducing atmosphere, tend to alleviate this constraint [49]. They have been successfully used to draw tungsten wire up to about 1.0 mm diameter where the wire temperature was in excess of 1270 K.

6.5 TOOLS FOR MINERAL EXPLORATION

To the mining engineer, there are basically four types of drilling: blast hole drilling, mineral exploration drilling, oil or gas exploration drilling, and oil or gas production drilling; the latter requires significantly larger boreholes to permit oil/gas flow. Exploration bore holes may extend from several dozen metres to several kilometres — 12 km is the current record — into the earth's crust. There is 'on-shore' (inland) drilling and offshore drilling from massive platforms. Most boreholes are vertical, but holes at an angle and even horizontal holes may be required depending on the circumstances.

The basic diamond drill tools include (Fig. 6.12):

— core bits,
— non-coring (or full hole) bits (crowns),
— reamers,
— stabilizers,
— casing shoes and other casing parts.

Diamond core bits are very widely used for the extraction of cylindrical rock samples, called cores. Non-coring or full hole bits are used to drill holes for cement grouting, explosives, etc. Ring-type reaming shells are generally placed above the bit at the core barrel base. They are used to stabilize the bit and the core barrel when running in the hole, slightly enlarge the hole to ensure that a new, full gauge bit will freely follow a worn bit after the latter has been withdrawn from the hole. Tapered reamers are used to enlarge the borehole when casing is required in the hole to seal-off air pockets or in unconsolidated formations. Coring operations can then be continued without normal diameter reduction. Round insert type stabilizers can replace the more expensive reaming shells in shallow drilling or in soft formations where the loss of bit gauge (O. D.) is minimal. Insert-type stabilizers have diamond impregnated tungsten inserts. Insert locations on the steel

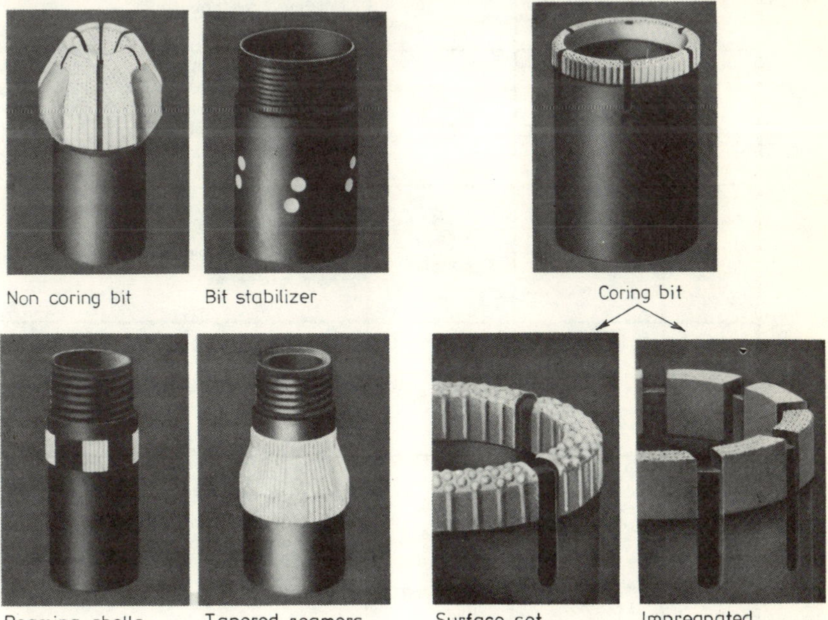

Fig. 6.12 — Selection of drill tool types for geological exploration (courtesy of Diamant Boart SA, Belgium).

Table 6.6 — Shapes of coring and non-coring bits [141]

SHAPE OF CORE BIT KERFS			
No.	Bit Contour	Code	Remarks
1	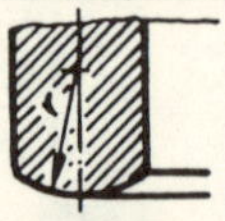	Standard Shape	Standard crown shape for DCDMA and metric core bits type A, B, T. Low carat weight. Recommended for medium hard rock, moderate abrasiveness.
2	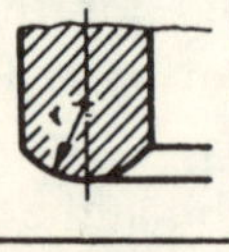	W-Shape	Rounded face allows stronger setting at the ID and OD. Little higher carat weight as for No. 1.
3	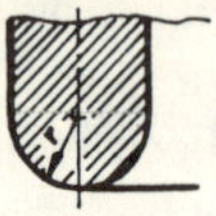	BY-Shape	Fully rounded bit shape, strong setting at gauges. Standard design for Large Serie bits and type K, Y and Z of metric standards. Suitable for hard and abrasive rock.
4		A-Shape	Unsymmetric bit shape. Penetration rate often a little higher than for W-Shape. Often used in USA for AX – NX and Wireline Series 10, 12 and Q.
5		Pilot Core Bit	Pilot bit with good stabilization in vertical holes. Penetration higher than with W-Shape. NX, K3, WL for medium hard formations.
6		Modified Pilot Bit	Modified pilot. This design is very often used in USA. For medium hard to hard formations.
7		2 Steps Core Bit	2-Steps core bit for thick kerf core bits. Good penetration, face weaker than for No. 1 – 6. For soft to medium hard formations.
8	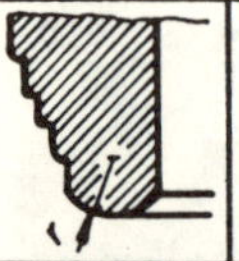	4 Steps Core Bit	Standard crown for wireline drilling in soft to medium hard formations. Good penetration and good stabilization. Too soft for hard formations.
9	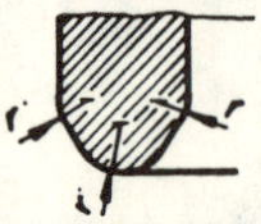	Double Cone	Compact bit, better stabilization than BY-Shape, suitable only for core bits with thick kerfs. Good penetration. For hard formations.
10	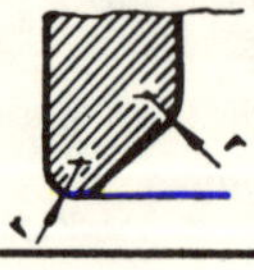	Special Inner Cone	Special inner cone. Suitable for core bits with thick kerfs only. This core bit is used with good performance also in Conglomerates.

Table 6.6 (cont.)

SHAPE OF NON-CORING BITS		
No.	Bit Contour	Remarks
1	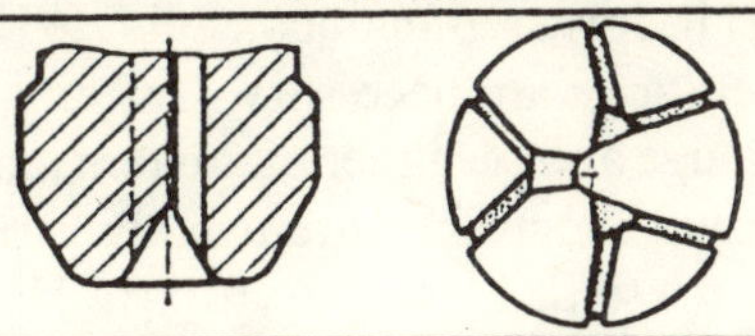	**Double Cone Shape** 3–6 waterways, diamond set pads are variable in wide ranges, bit has good stability, little direction tendency.
2	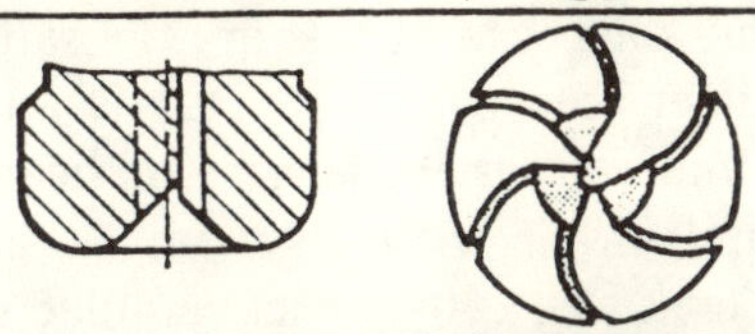	**B-Shape** Spiral waterways secure good flushing, suitable for hard rock, good penetration rate, slight tendency to deviate.
3	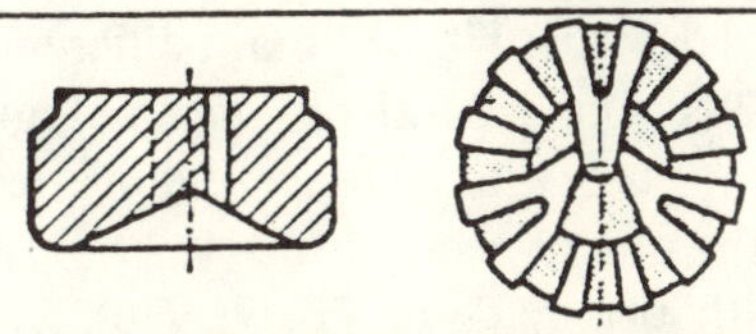	**Wide Inner Cone** Light setting, many waterways and small diamond set pads, bit needs relatively little load, recommended for soft to medium hard formations.
4	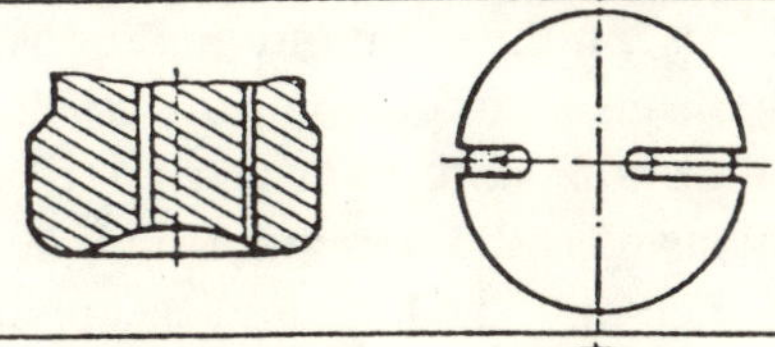	**Concave Bit** In Europe not very often used. Flushing of center not favourable.
5	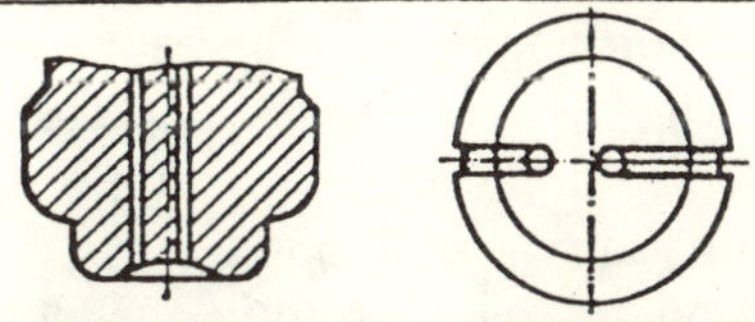	**Pilot-Concave Bit** Similar to No. 4. Relatively weak center, not recommended for hard or abrasive rock.
6	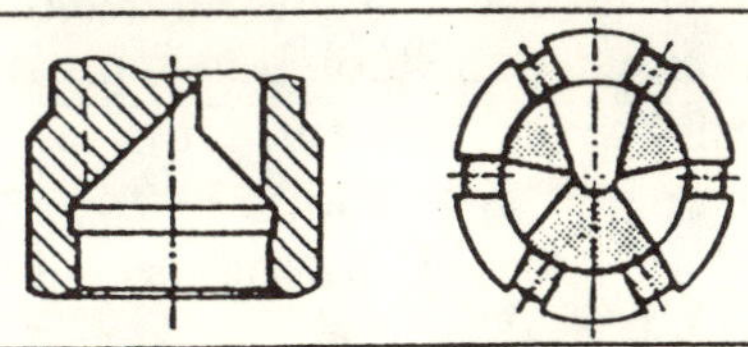	**Deep-located Inner Cone** Extremely long reaming edge, excellent deviation control under high bit loads. Not recomménded for formations with a tendency to break easily.
7	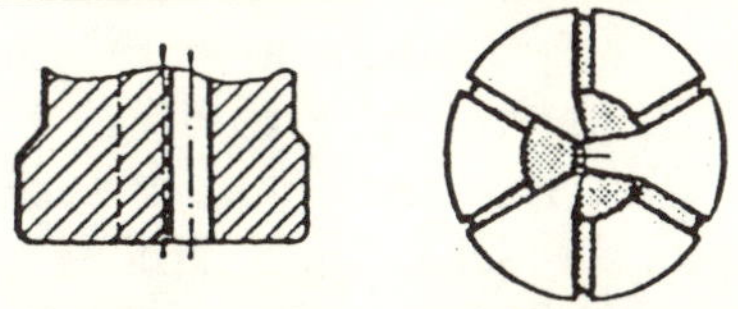	**"Flat Bottom"** Special bit to flatten bottom of the hole, for instance for positioning of instruments.
8	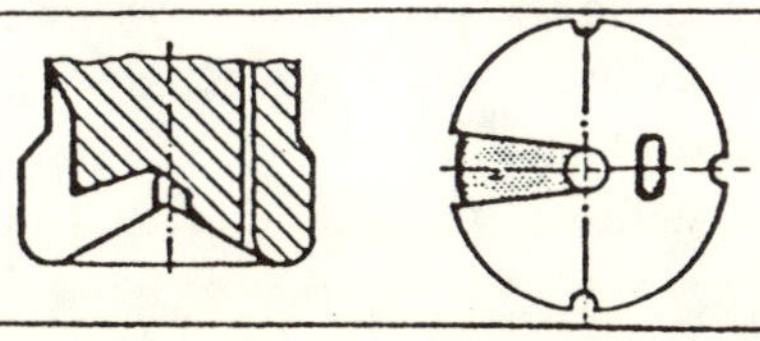	**"Core Splitter"** Bit center is lifted upwards allowing to break the little core.

body are predetermined in all cases, thus providing for optimum tool life and stability. The difference between a casing shoe and a casing bit lies in the inner set diameter. The casing shoe has no diamonds set on the inner gauge and has an I. D. that will allow the corresponding core barrel, bit and shell to pass through it. It is used in unconsolidated formations and where rotary casing operations are necessary. Casing bits are set with diamonds on the inner gauge and the inner diameter is smaller than that of the corresponding casing shoe. They are usually used when drilling hard formations prior to setting of the casing. The O. D. of the corresponding casing shoes and casing bits are the same [63, 141].

The basic drill tool dimensions, such as O. D., I. D. and screw threads are standardized, in the Imperial system — e.g. American DCDMA and Canadian CDDA — and in the metric system, as adopted e.g. by Diamant Boart companies as well as by Soviet manufacturers. There is a great diversity of drill tool designs, with the main differences in the shape of the rim or crown. This will depend on the geological conditions known or forecast where the tool is to be used, and in selecting a drill bit, account is also taken of the rate of penetration (R. O. P.) and the drilling parameters available with the machine (Table 6.6). The R. O. P. will depend, among other things, on the diamond quality used in the tools.

The three basic mechanisms of rock drilling are crushing, by pressing e.g. a diamond into the rock, abrasion, and shearing (Fig. 6.13). In most cases one of these mechanisms is the dominant one, thus deciding the main tool specifications, including the kind and size of diamond. In the case of surface-set tools, the drilling process can be thought of as plastic deformation of the rock, which leads to its gradual crushing. Such tools are set with relatively large natural diamonds of appropriate shape (see Chapter 3) or with carbonado-type diamonds.

The harder the rock and, more importantly, the more abrasive it is, the finer should be the diamond in the tool. Bits impregnated with high thermally and mechanically shock resistant abrasives, especially synthetic diamond grit ranging in size from 250 to 800 μm, are becoming increasingly popular. The rock drilling mechanism involved in the case of impregnated tools is abrasion.

Typical surface speeds of surface-set diamond tools are up to about 3 m/s, while impregnated bits are run at 2 to 5 m/s [141]. Synthetic diamond bits offer longer life at faster and more consistent penetration rates. They can be used over a wider range of drilling speeds and with greater thrust and impact loads. This is because synthetic diamond can be grown to a regular, blocky shape which is mechanically stronger than a crushed natural diamond particle.

Drill tools armoured with thermally stable PCD inserts such as

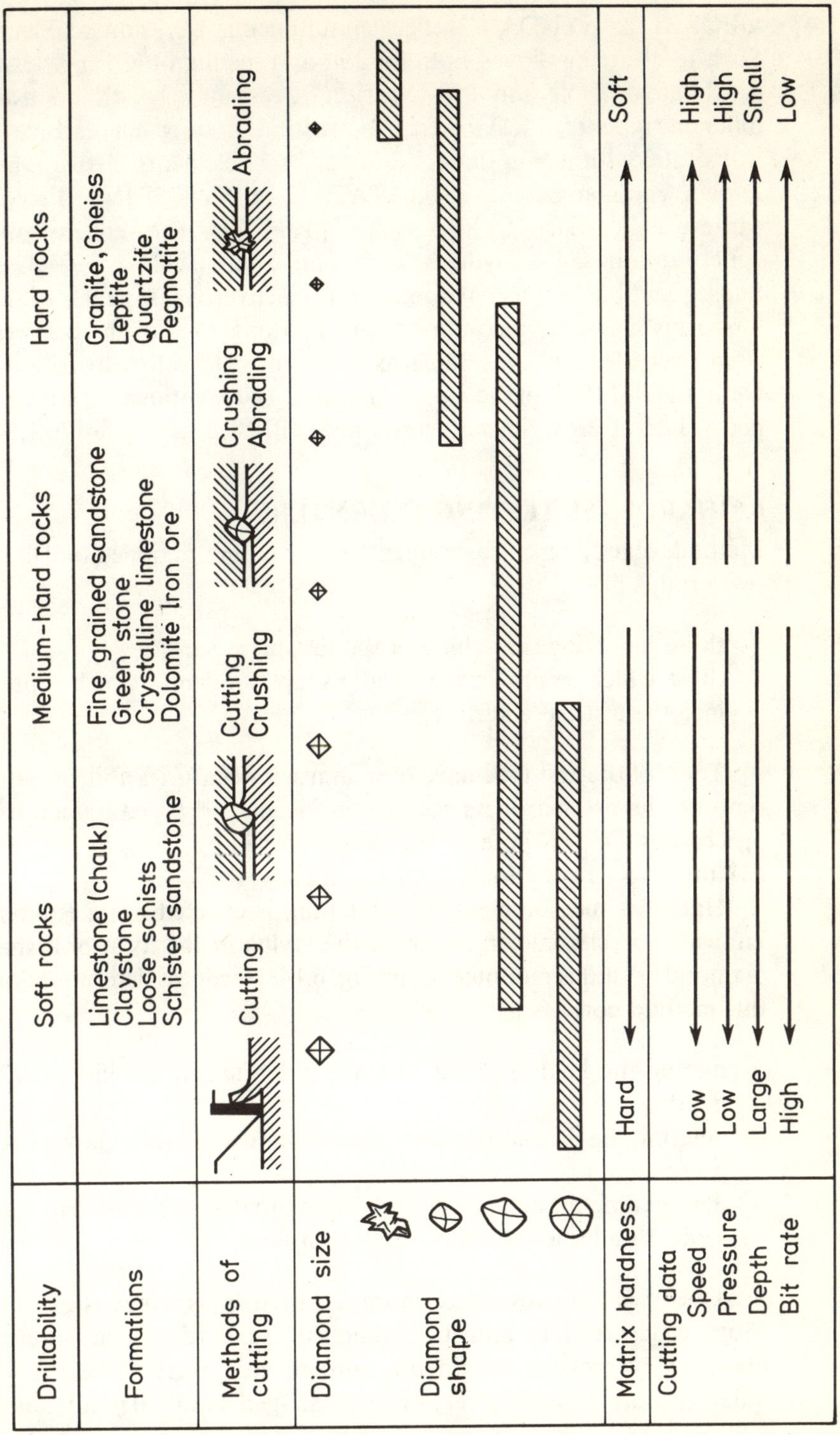

Drillability	Soft rocks	Medium-hard rocks	Hard rocks
Formations	Limestone (chalk) Claystone Loose schists Schisted sandstone	Fine grained sandstone Green stone Crystalline limestone Dolomite Iron ore	Granite, Gneiss Leptite Quartzite Pegmatite
Methods of cutting	Cutting	Cutting Crushing / Crushing Abrading	Abrading
Diamond size			
Diamond shape			
Matrix hardness	Hard ←		→ Soft
Cutting data Speed Pressure Depth Bit rate	 Low ← Low ← Large ← High ←		 → High → High → Small → Low

Fig. 6.13 — Method of rock drilling, its hardness and the recommended kind of drilling tool and diamond [169].

GEOSET or SYNDAX-3 effect simultaneous abrasion and cutting. Such tools are employed both for soft and medium-hard rock drilling.

Relatively soft and low-abrasion rocks may be drilled using a different type of PCD insert, comprising a tungsten carbide stud (substrate) with a thin diamond layer. PCD elements of this type are known commercially as STRATAPAX or SYNDRILL. From the earliest field trials, STRATAPAX [191] bits showed exceptional penetration in salt, anhydrites, carbonates, marls, clay and sandstones. Subsequent bit designs have consistently delivered fast penetration and long runs in soft to medium-hard formations. One of the best results from over 350 runs was 386.8 m at 6.7 m/hr in salt/anhydrite in the Netherlands [191]. In the same formation, a conventional roller cone bit penetrated at 2.7 m/hr, and could not drill effectively after 70.1 m.

6.6 HARDNESS TESTING INDENTERS

Methods of hardness measurement using a diamond tipped tool fall into two groups [196]:

— those involving scratching of the test piece surface,
— those which employ a specifically shaped indenter which is pressed into the test piece surface.

Tools of the first type have been in use for centuries and, in the 18th century, the first hardness scales were developed based on scratching procedures. One of those scales was developed by Mohs in 1812 and it continues to be commonly used to this day.

Hardness measurement by scratching is effected using equipment known as a sclerometer, in which the stylus in the form of a shaped diamond crystal is mounted, as in Fig. 6.14. Hardness determination by this method consists in:

— finding the load required to make a scratch of specified width or depth, or
— determining the width of the scratch obtained under a known cutting point load, or
— determining the magnitude of the tangential force required to make a scratch under a known cutting point load.

Thus it is wear resistance more than hardness which is measured. More widespread in industrial practice are hardness measurements obtained by pressing an indenter into the test piece surface. Typical indenter tools consist of a very precisely shaped diamond point mounted

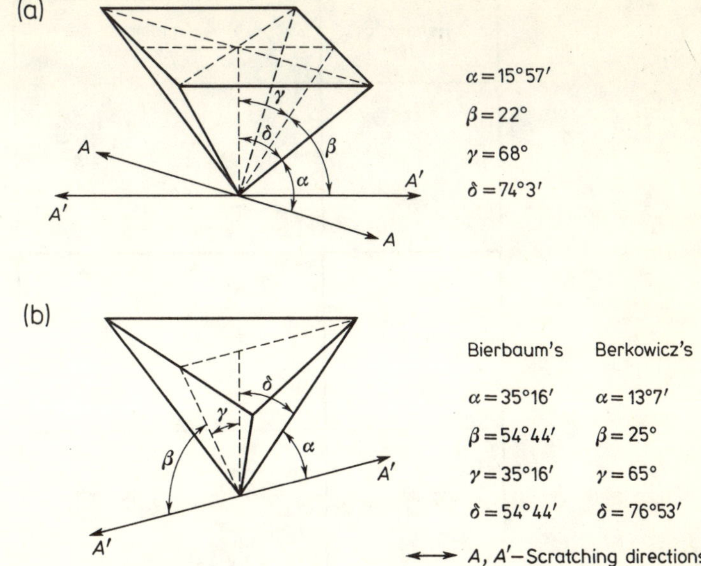

Fig.6.14 — Diamond crystals employed in scratching hardness measurements [139]: (a) tetrahedral Vickers pyramid; (b) Bierbaum's and Berkowicz's trihedral pyramids for scratching directions.

in a holder. Measurements of the dimensions of the indentation are made under a microscope, and the load exerted on the indenter is recorded to high precision. A hardness number is obtained based on load over area, expressed in kg/mm^2. The most common scales are the Rockwell (a cone shape) and Vickers (tetragonal pyramid indenter shape). For microhardness testing, the indenter developed by Knoop is widely used where the pyramid-shaped tip is elongated along one of its axes (Fig. 6.15). The advantage of a Knoop indenter is the relatively high accuracy of measurement thanks to the use of a longer measuring segment. On the other hand, the Knoop indenter does not make it possible to observe the anisotropy in hardness of a given material, or calculate the effects of directional stresses such as may be observed in tests with e.g. a Vickers indenter. Other diamond tip designs, such as those of Grodziński (double cone), Berkovich (trihedral pyramid), or modifications of the Vickers pyramid, are used only very infrequently [196].

Only the highest quality single crystal natural diamonds can be used to make indenters. Moreover, the crystal must be cut precisely enough for its greatest hardness and abrasion resistance directions to coincide with the indentation direction. For example, a Rockwell indenter tip should have the shape of a simple cone with a rounded vortex. The apex angle is $120° \pm 30'$ and its axis should coincide with the tool shank axis to

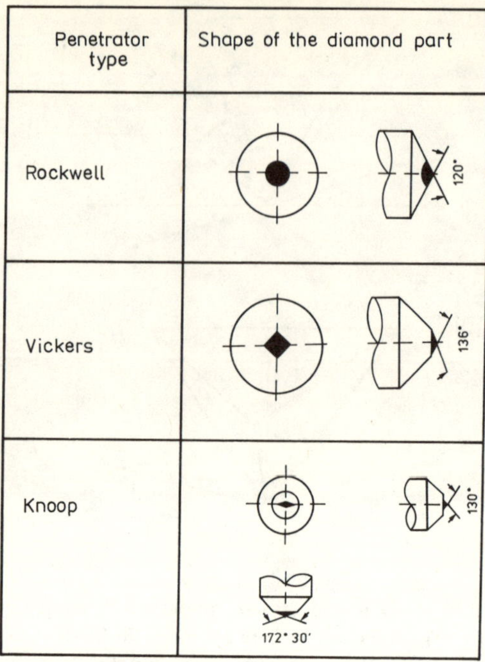

Fig. 6.15 — Diamond penetrators for hardness measurements.

an accuracy of 0.5°. The tip radius is 0.2±0.01 mm and the vortex contour should not deviate from the nominal contour by more than ±0.002 mm. The tip is used for making indentations in materials of hardness above HB 200 (Brinell scale). On the basis of the indentation measurements and load, the hardness is expressed on either the Rockwell C or A scales [208]. The crystals used in these tools range from 0.1 to 0.4 ct. As already noted, the Vickers indenter has the shape of a regular pyramid with a square base. The angle between the opposite pyramid walls should be 136°±20′ and between the opposite edges — 148° 7′. At the same time, all four walls of the pyramid should be inclined at the same angle to its axis (to an accuracy of 30′), and the vortex should be pointed (sharp): the admissible length of the vortex edge made by opposite walls should not exceed 0.002 mm [139]. The indenter surfaces should be well polished and show no cracks, chips, scratches, or other surface defects. The state of an indenter tip is checked visually from time to time under at least 50 × magnification.

6.7 DIAMOND STYLI FOR SURFACE FINISH MEASUREMENTS AND DIMENSIONAL GAUGING

Very high hardness and abrasion resistance make diamond particularly suitable as the material from which to make gauging points for dimension and quality checking of work-pieces. Naturally enough, the diamond points are set in the place(s) which make(s) contact with the work-piece. Diamond gauging instruments may be designed for manual use, e.g. micrometers with diamond tips, or they may be built into equipment in which case their operation is automatically controlled.

The design of surface finish testing equipment will be determined by the work-piece(s) being investigated, both in terms of their shape and dimensions. Equipment is usually built on a lever principle with a diamond tip at one end (Fig. 6.16), the other end of the lever being linked

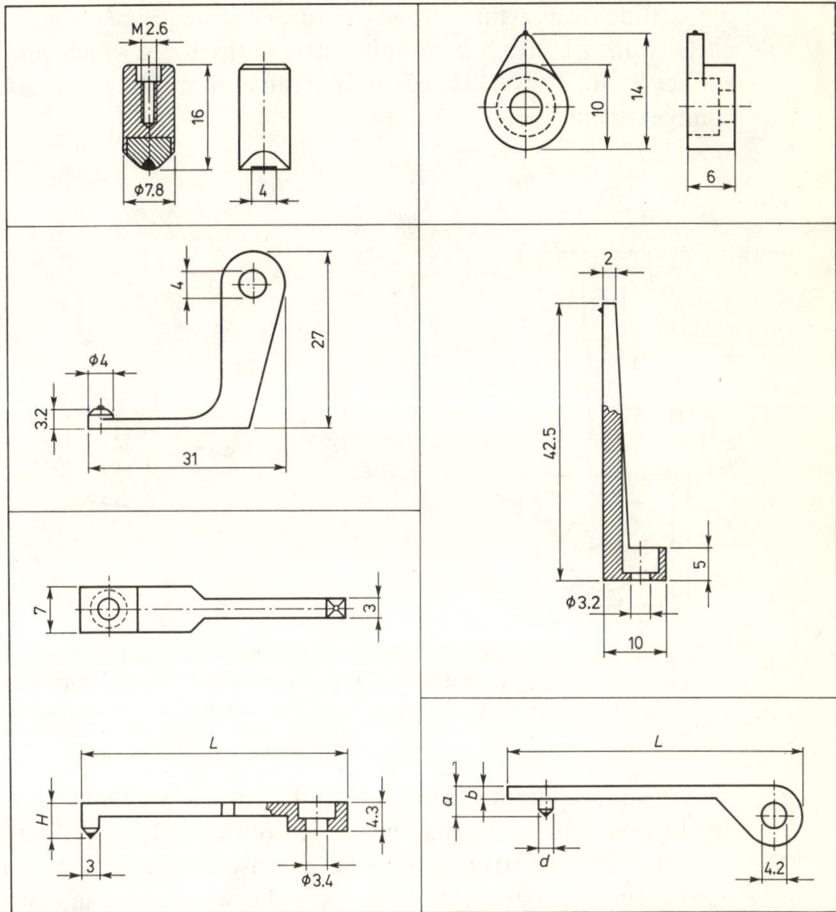

Fig. 6.16 — Selection of contact tip designs for linear measurements [139].

to a mechanical, optical or electronic sensor for measuring the movements of the lever. As the diamond tip passes along the object's surface, any surface irregularities and deviations from the expected profile cause a deviation or vibration of the lever and these vibrations are recorded, usually in paper chart form. To minimize stylus wear and scratching of the work-piece surface, the diamond tip is usually spherical. Diamond tips are polished in such a way that gauging pressures of up to 20 kN do not produce any discernible traces. The diamonds usually employed are from 0.005 to 0.3 ct. Diamond feeler styli, according to De Beers [72], are to be found in most precision engineering workshops or research laboratories as the points of contact in high precision instruments which measure surface finish. The stylus tip is usually cone shaped and has a radius as small as 0.0012 mm on a high resolution stylus. Surface measuring devices such as the Talysurf or Profilometer find wide application where fine surface finishes are required. They not only check work quality but can also inspect the tools which produce the surface finish. They can check the overall accuracy of plain, curved, and complex multi-curved surfaces.

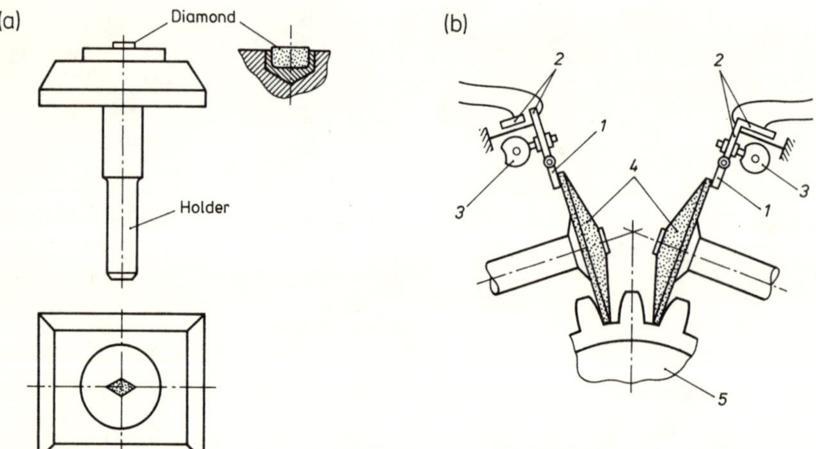

Fig. 6.17 — Maag type contact tips (a) and schematic diagram of the arrangement in which they are employed (b) [139]: *1* — lever with diamond compensation bit, *2* — electric contacts, *3* — cam, *4* — grinding wheels, *5* — toothed wheel being ground.

Linear measurements and, especially, in-process control under mass production conditions, may be carried out directly or indirectly using diamond-tipped instruments. In the first case the diamond gauging point makes semi-permanent contact with the work-piece under test and continuously monitors its dimensions. The moment the specified dimensions have been attained, the mechanical or electronic system linked to the diamond point sends a signal to the tool control system to discontinue machining. In instruments based on indirect measurement methods, the diamond does not make direct contact with the work surface but, instead, acts as a kind of stop on the tool head feed movements, e.g. constraining the infeed of a wheel grinding a work-piece. The desired range of tool movements is set very precisely before machining commences. Diamond distance stops minimize inspection time during such operations as lapping or honing, when the finished job is of predetermined size. One of the most common uses of diamond in measuring instruments of this kind are the so-called compensation tips for Maag-type gear grinders (Fig. 6.17). Such contact tips maintain a fixed distance between the end faces of the grinding wheels.

6.8 DIAMOND TOOLS FOR BURNISHING

An entirely different aim is achieved with diamond-tipped tools which deform surfaces by cold-working. Pressure induces plastic deformation of the work-piece surface layer [139]. Such tools are used for burnishing (or planishing) the surfaces of, especially, non-ferrous metal and steel work-pieces having suitable plastic properties. Cold-working may sometimes constitute a preferable substitute for polishing, lapping,

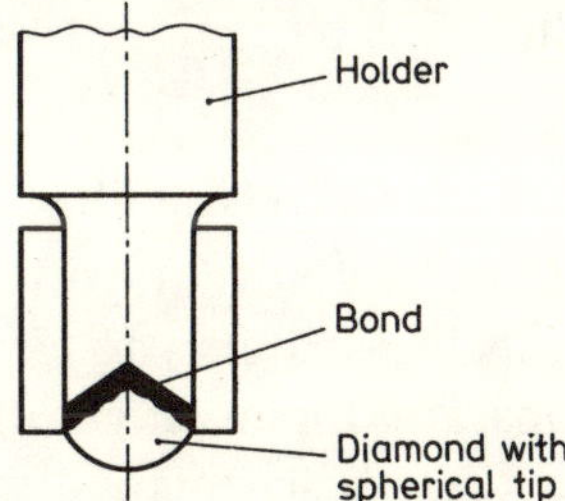

Fig. 6.18 — Example of tool design for cold working.

honing, etc. Burnishing tools are fairly similar — in terms of their design — to other single crystal or PCD tipped tools, except that instead of defined tool edges the tip is spherical and highly polished (Fig. 6.18). In operation, a spring-loaded holder urges the diamond sphere onto the work-piece surface with a predetermined force.

7

Non-cutting uses of diamond

While the different tools described so far in this book represent the most common industrially important diamond products, they do not exhaust the possible applications of this precious mineral. There exists, in other words, quite a wide range of other diamond products which are, however, less well known because of their highly specialized uses. As such they do, nevertheless, play an important role in various fields of science and technology.

7.1 DIAMONDS IN HIGH-PRESSURE CELLS

A good example of how advantage is taken of the great strength of diamond and of its transparency to electromagnetic radiation is in the anvils used in apparatus for investigating changes occurring in materials under high pressures. A schematic diagram of such an apparatus is shown in Fig. 7.1. In spite of its simplicity, the device is capable of generating pressures of up to 20 GPa in the space between the precisely profiled diamond pistons. In addition, the apparatus can be extended to permit heating of the sample. The wide range optical transparency of diamond permits direct observation of the physico-chemical transformations taking place in the substance placed between the diamonds. Specially shaped and polished diamonds are used as windows for this sort of work, in pressure cells for visual examination of test materials with the aid of a microscope, and as anvils in high pressure X-ray cameras. They may also be used with infra-red or ultra-violet spectrometers to ascertain absorption characteristics of materials under pressure [50, 181].

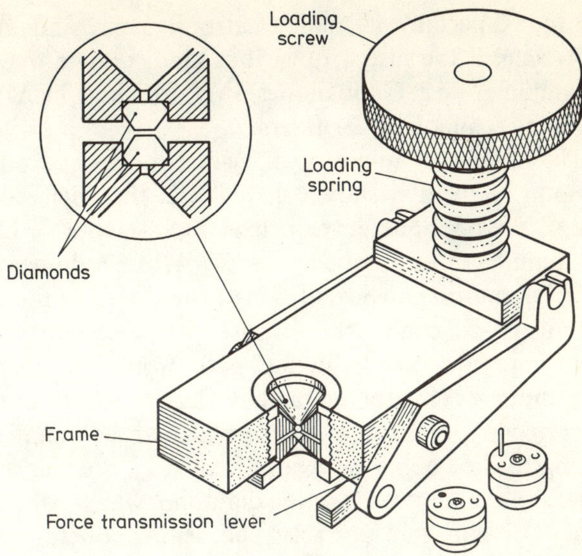

Fig. 7.1 — Diamond cell for investigating transformations under high pressures [80].

7.2 DIAMOND ELEMENTS IN ELECTRONICS

Diamond's unique combination of physico-chemical properties is exploited in some kinds of electronic devices. Thus, appropriately prepared diamonds are commonly used:

— to extract heat from a given area of an electronic device or element,
— as ionizing radiation detectors,
— in temperature measurements,
— in the manufacture of abrasion-resistant parts.

Diamond elements for the above applications are cut and polished from specially selected high quality natural crystals. In recent years, a great deal of research has been conducted on synthetic crystals obtained by controlled growth under very high pressures and temperatures, and by epitaxial methods. Diamonds for electronic applications are exceptionally thoroughly tested using highly specialized equipment, every single crystal being tested for the property which is basic to its application in the given electronic device.

As is well known, heat emission accompanying current flow may be harmful for electronic devices and equipment in that excessive heating may affect the electric characteristics. Thus rapid and continuous cooling of such elements and extraction of heat away from them is often essential, especially if small electronic parts and large currents are involved, as is the case with various kinds of high-frequency generators and amplifiers, e.g. Gunn diodes, IMPATT and TRAPATT diodes, etc., as well as semiconductor lasers.

The design of microwave oscillator diodes and the position of diamond in them is shown in Fig. 7.2, the device depicted being an IMPATT type diode developed at Bell Telephone Laboratories [72]. The manufacturing sequence for TRAPATT diodes is represented in Fig. 7.3. According to Seal [179–181], in diodes of this type the diamond heat sink is gold coated, the purpose of the metallization being threefold: one, it provides a conducting path from the lower surface of the semiconductor chip (or 'epitaxial wafer') to the lower electrical contact, the copper base. Two, it also provides an adherent metal surface for brazing to the copper base, and finally, it provides a clean, soft metal surface on the upper side of the diamond, which will stick when pressed onto a similar gold surface. The semiconductor chip can then be mounted in position by a gold-to-gold 'thermal compression bond'. The complete unit is an oscillator and it will generate radio waves of very high frequency when a suitable voltage is applied across the two

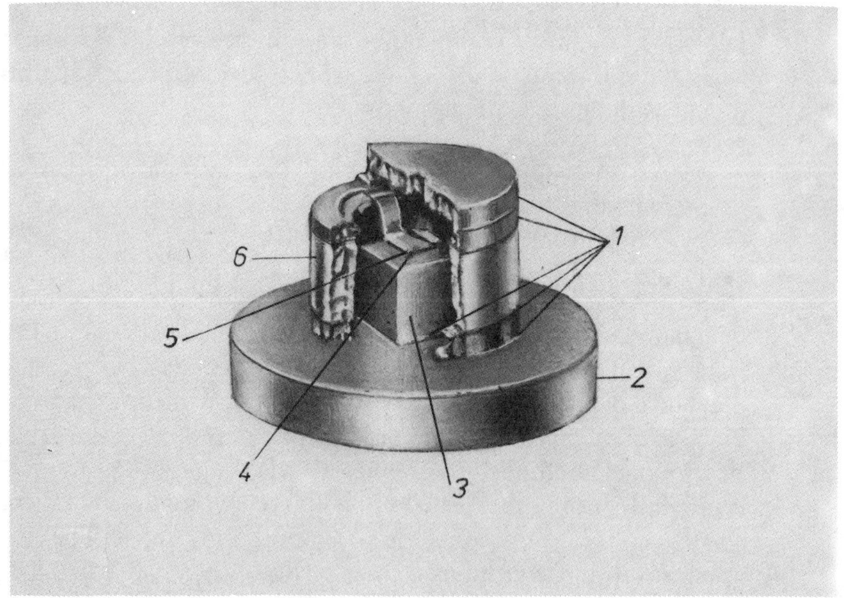

Fig. 7.2 — The construction of a microwave oscillator diode with diamond heat sink [72]: *1* — brazed joints, *2* — copper base, *3* — gold-coated diamond, *4* — gold-to-gold thermo-compression bond, *5* — epitaxial wafer, *6* — microwave ceramic-metal package.

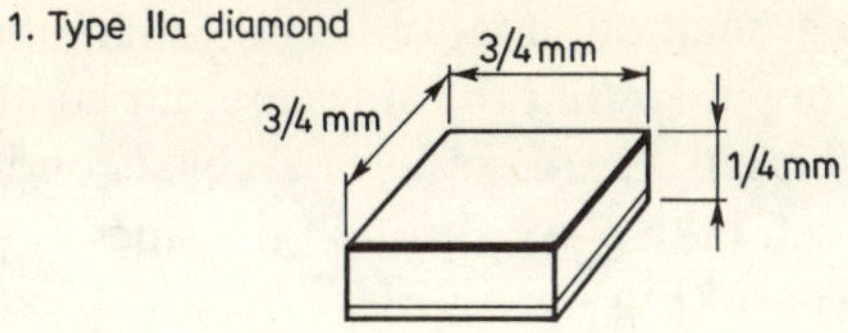

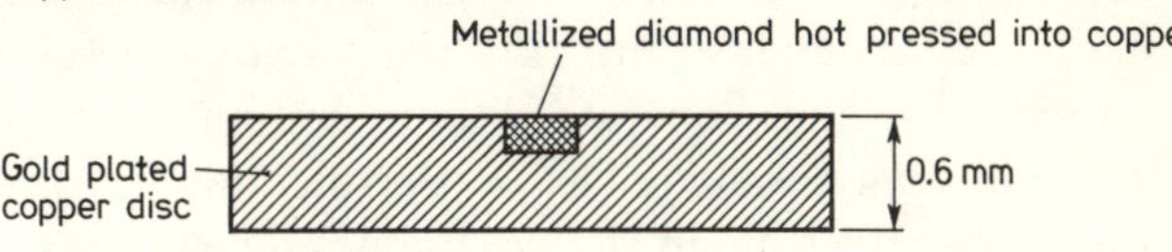

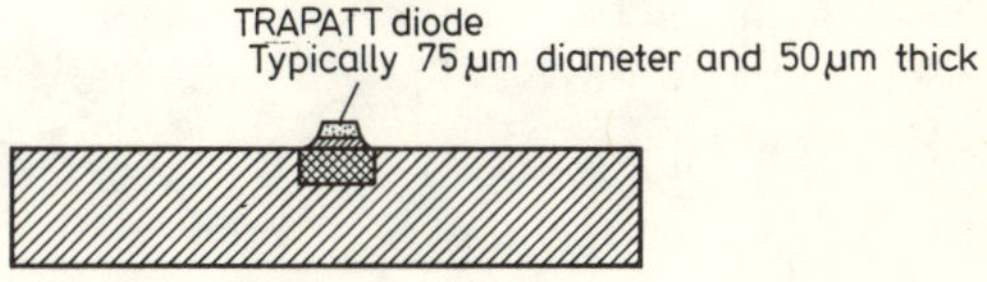

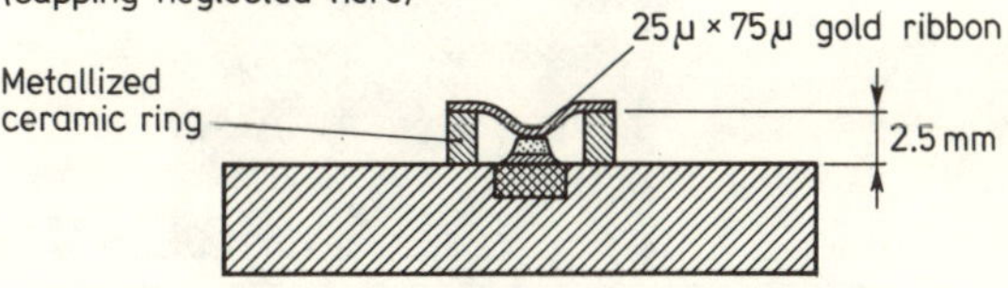

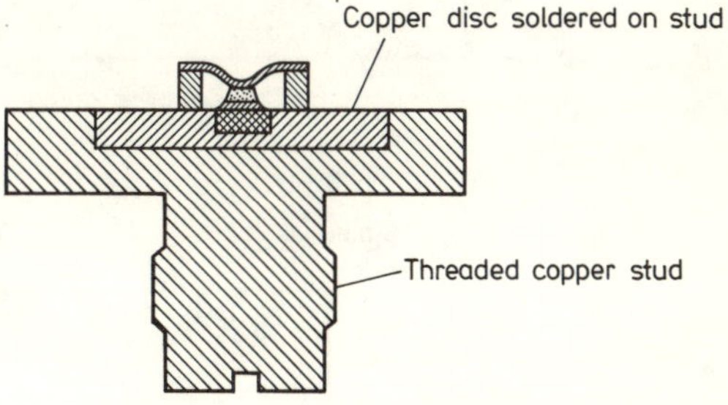

Fig. 7.3 — Stages in the fabrication of a TRAPATT diode with embedded diamond heat sink [72].

terminals. Normally it is mounted in a cavity in a metal block and feeds the radio waves to a connecting waveguide and thence to the antenna. The diamond heat sink does not absorb heat but transmits it rapidly to a relatively massive piece of copper, from which the heat passes to its surroundings. Another method of mounting the diamond is to embed it in a larger mass of copper so that the sides are buried and only the top surface is exposed. This has the advantage that heat conduction from the diamond to copper can take place through the sides of the diamond as well as through the base (Fig. 7.4).

A semiconductor laser is a multilayer structure in which laser light generation occurs as a result of electron pumping from a stripe source. Heat is also generated and there is a need for efficient cooling of the junction. Diamond heat sinks are again an obvious choice, as the devices are small and should operate with maximum reliability and at as low a temperature as possible. The first semiconductor lasers to operate at room temperature did in fact use diamond heat sinks — previously they had required cooling with liquid nitrogen. It is now possible to operate

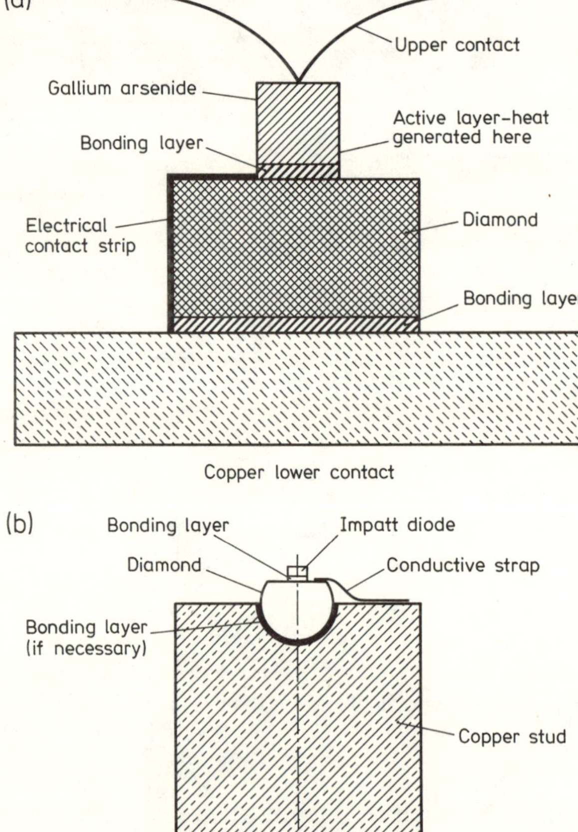

Fig. 7.4 — Schematic diagrams of a Gunn diode on (a) a rectangular diamond heat sink (Seal, 1971) and (b) a spherical diamond heat sink [80].

semiconductor lasers at room temperature without diamond heat sinks, but the use of diamond does give better reliability and some advantages in fabrication, and has—according to one pair of authors—become 'traditional with room temperature, continuously operative communication-type lasers'. On the other hand, the high thermal impedance of the gallium arsenide laser structure itself is sufficient in many cases to nullify most of the thermal advantage of the diamond, and the advantages then lie mainly in the wear resistance and stiffness of the diamond, in the high quality of polished diamond surfaces, and in the very precise tolerances to which diamond components can be made [72].

When diamond is used in radiation detectors, advantage is taken of the change in the optical and electrical parameters of the crystals when they are exposed to ionizing radiation [72, 80]. If a high external electric field is applied across a diamond, a conductive pulse may be measured during bombardment with ionizing radiation. The counting properties of natural diamond appear to be very type-dependent. Most Type IIa diamonds, classified as such by ultra-violet light transmission, are good counters, and exhibit a saturation maximum pulse height for a field of more than about 30 kV/cm [80]. Type I diamonds, however, will only count if the applied field is greater than 50 kV/cm.

Vermeulen [211] has developed a process whereby Type IIb semiconducting diamonds can be successfully used as radiation detectors and counters. Type IIb diamond is rendered non-conducting by the creation of a thin band of radiation damage, using high energy electrons. If a voltage is applied across the diamond, the thin band must extend over the entire current flowing area. With the voltage applied, current will pass through the diamond only when the ionizing radiation is incident on the diamond. Untreated Type IIb semiconducting diamonds are not suitable for use as conduction counters.

Another effect of which advantage can be taken in diamond radiation detectors is fluorescence (Section 1.4.2). The variation range in the scintillation efficiency is small compared to the variation in other electrical and optical properties. Thus, diamond can be used as a scintillation counter. Because of the relatively weak intensity of the light flashes generated by ionizing particles, a diamond detector is optically linked with a photomultiplier which converts light pulses into electrical pulses and amplifies them [80].

The high thermal conductivity and low specific heat, high hardness and abrasion resistance as well as chemical passivity make diamond particularly suitable for the production of electronic devices for temperature measurements, exemplified in Fig. 7.5 by diamond thermistors. These can be made to detect changes in temperature as small as 0.002 K.

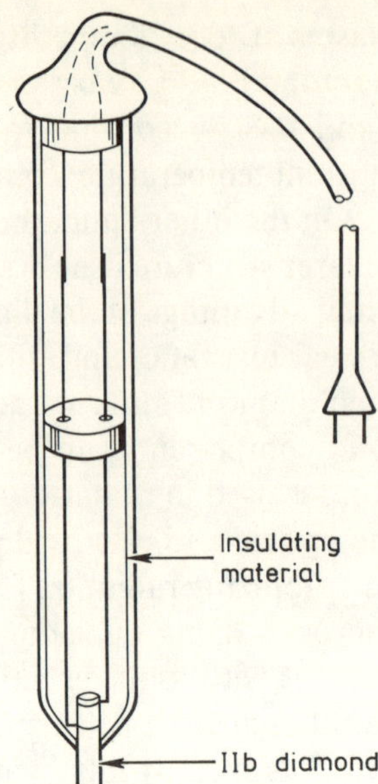

Fig. 7.5 — Schematic diagram of diamond temperature gauge.

Use is made here of e.g. cylindrical shapes cut from Type IIb diamonds, the surfaces of which are metal coated and linked to contact leads. Also synthetic diamonds doped with boron can be used for the purpose. Rodgers and Raal [80] developed a method of fixing platinum leads to semiconducting diamond cylinders. A titanium-silver-copper alloy is used to bond the platinum to the diamond.

The extremely high abrasion resistance of diamond is also exploited in the manufacture of so-called 'mechatronic' electronic elements which are susceptible to wear. A good example is the gramophone stylus, the tip of which should ideally maintain its shape and size for ever. In practice, using a well-designed, balanced pick-up arm, a sapphire stylus lasts about 25 to 30 playing hours before wear begins to damage the record, and a diamond stylus over 600 playing hours [72]. Despite the introduction of automatic equipment, the manufacture of diamond gramophone styli calls for considerable engineering skill. A typical finished stylus is only about 0.675 mm long and 0.375 mm in diameter, yet such is the efficiency of the diamond impregnated wheels used to grind them that one operator can produce up to 50 styli an hour. The diamond tip is ground into a cone shape and has a radius polished on the tip — the dimensions depending on the size of the record grooves. A tip with a 0.0875 mm radius would be used on the now-obsolete 78 r.p.m. shellac records. LPs require a tip-radius of 0.0175 mm and for very high quality stereo reproduction, an elliptical stylus with a radius of only 0.005 mm at the point of contact is used. The laser-scanned 'Compact Disc' will have an obvious effect on this market.

Yet another application area involves a thin layer of diamond deposited from the gas phase [101]. Such diamond layers may be vapour deposited onto silicon or metal parts. The electrical properties of such layers depend on the conditions of their manufacture, and their conductivity can be varied over a wide range [101, 188]. A big future is predicted for diamond films in applications as diverse as space rocket windows and razor blades, according to a US newspaper report.

7.3 DIAMOND OPTICAL ELEMENTS

Because of its transparency to electromagnetic radiation, diamond is used in various kinds of optical elements. Thus, small 'windows' are made from diamond crystals for all manner of research equipment,

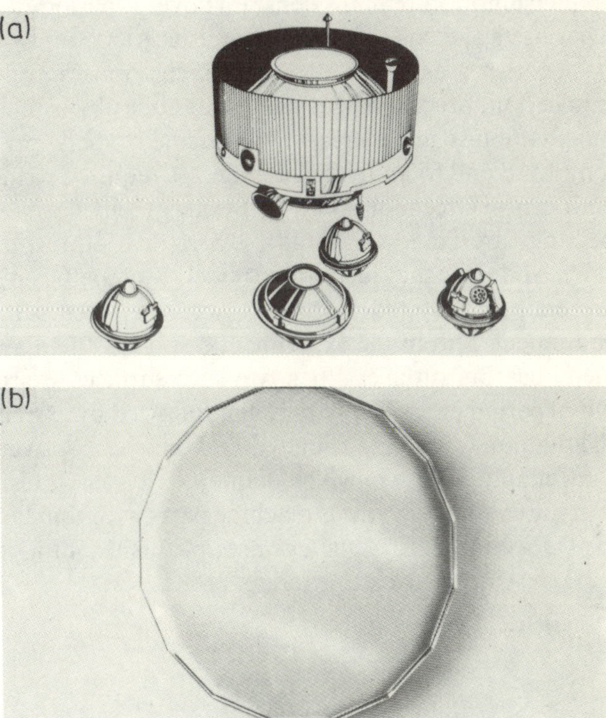

Fig. 7.6 — Use of diamonds in space research: (a) Pioneer Venus multiprobe — the spacecraft, launched as a single vehicle, was split into five probes some 11 million km from Venus. Four of the probes carried Drukker diamond optical components; (b) the largest diamond window on Pioneer Venus probe (diameter 18.2 mm, thickness 2.8 mm) (courtesy Drukker and Zn, Amsterdam).

including that used in space probes [24, 70]. Diamond lenses and prisms are occasionally made from good quality diamond where a particularly high refractive index is essential [179–181]. One company noted for the production of diamond optical elements is D. Drukker and Zn of Amsterdam, Netherlands (Fig. 7.6).

7.4 DIAMOND WEAR PARTS

The outstanding abrasion resistance of diamond is put to good use in the production of all kinds of fuel metering or dosimeter nozzles. These measure the fluid flow per unit time. Diamond nozzles are also employed in some central heating installations [80, 198]. Heating systems in some parts of the USA have been fitted with automatic oil burners incorporating diamond nozzles. With nozzles made of other materials, the harsh, abrasive action of the oil mist causes a slow but constant increase in oil consumption as the bores become worn. Using finely drilled diamonds, oil consumption remains constant: over a period of time, the saving in fuel offsets the cost of the diamond insert.

Another area where diamond's high abrasion resistance is put to use is in certain precision bearings. Synthetic ruby is most commonly used, but diamonds are increasingly specified where particularly accurate, long-lived and friction free bearings are required to meet the demands of 20th century technology [80]. Besides their wear resistance, diamond bearings are far less affected by corrosive atmospheres or acids than ruby. And since they have an extremely low coefficient of expansion, they are eminently suited for use in accurately temperature-compensated equipment. Still in the experimental stage is the diamond impregnated metal bearing surface. When two such surfaces are rubbed together, an initial period of very high friction is followed by extremely low friction as the diamonds become rounded and slide easily over each other [80].

Yet another use in which diamond's abrasion resistance is exploited is in the coating of various machine parts with thin layers of diamond by CVD (Section 2.2.5). One example of this use of a diamond film is the work rests in centreless grinders [188].

8

Diamond jewellery

Although aesthetic tastes are relative and differ from one person to the next, diamond continues to be a major gemstone since antiquity. There exists some kind of myth of diamond as the symbol of beauty and wealth. The acquisition of unmounted cut diamonds and diamond jewellery has always been a way to demonstrate wealth and power.

Because of the high price and rarity, especially of large gem diamonds, many have been cut to special commission by royalty and the nobility. Some of these are regarded as outstanding, very precious works of art. Not infrequently they are objects of great historical value, as well as being perhaps the highest class example of material culture, collector's pieces gracing royal treasuries and the most famous museums, e.g. the Kremlin, the Louvre, the Tower of London.

The brilliance and fire related to light dispersion usually appear in the crystals only after the stone has been cut. The oldest cut diamonds come from ancient India. Until the end of the 15th century, diamond working was limited to polishing the natural crystal faces, with very little transformation of the original stone shape. Thence diamond crystals came to be cleaved and cut, then ground and the newly formed faces polished to make facets.

A large number of magnificent and variously shaped brilliants are now in existence. The recognized diamond cuts fall into four basic groups [27] : the brilliant cut, rose cut, emerald cut and marquise cut. In turn, each of these groups falls into a large number of specific varieties with their own, historically motivated names (Fig. 8.1). The evolution of a particular cut involves first of all changes in the number, shape and size of the facets. This in turn invokes changes in the number of symmetry

Fig. 8.1 — Some of the principal geometric shapes into which most precious and semi-precious gems are cut (after *Comprehensive Faceting Instructions*, D. L. Hoffman) [170] (marquise or navette cut, pear-shape cut, heart-shape cut, boat-shape cut, pear-shape rose cut, Brazilian cut, Lisbon cut, Gull-Holland rose

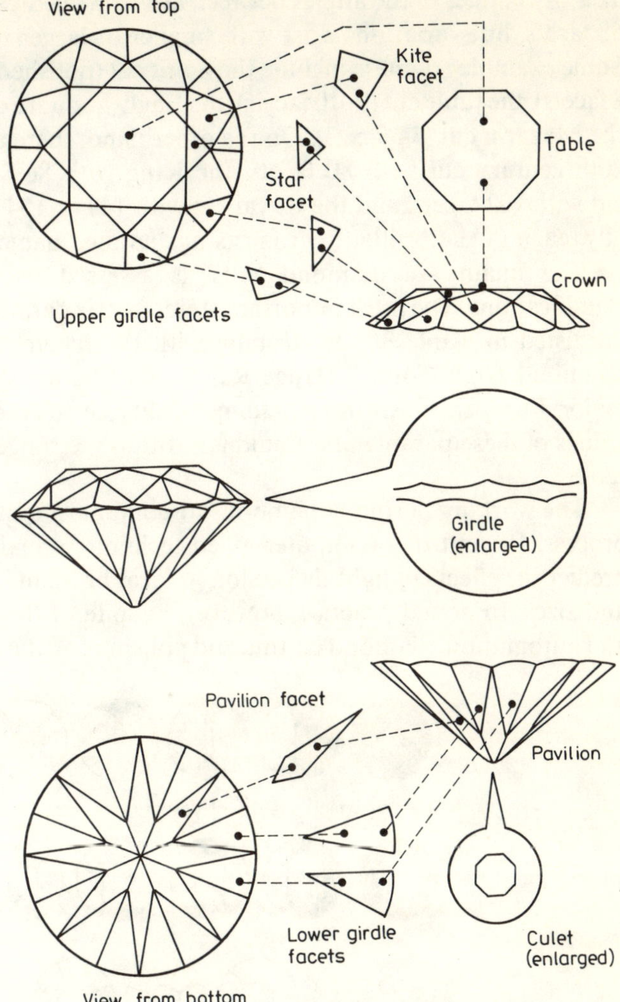

Fig. 8.2 — Names of facets on brilliants [27].

---

cut, double-rose cut, twentieth-century cut, jubilee cut, English square-cut brilliant, seminavette, step-cut bead, king cut, magna cut, old European cut, old-mine cut, single cut, Swiss cut, oval cut, trap brilliant cut, double-cut brilliant, three-facet rose cut, six-facet rose cut, Split-brilliant cut, English round cut brilliant, bullet cut, keystone cut, square cut, French cut, half-moon cut, rondelle, epaulette cut, tapered pentagon cut, kite cut, lozenge cut, trapeze cut, pentagon cut, calf's-head cut, window cut, hexagon cut, long hexagon cut, fan-shape cut, shield cut, whistle cut, rhomboid cut, square emerald cut, cut-corner triangle cut, long octagon cut, baguette, tapered baguette, emerald cut, table cut, bevel cut.

axes or changes in the angles between facets. A classical brilliant cut has 58 facets, but variations exist with smaller or larger numbers of facets. Some examples of these include the point cut (polished octahedron with 8 facets), the table cut (9–10 facets), the English square cut (30 facets), and the Mazarin cut (34 facets). On the other hand, there are the Jubilee, or 20th century cut with 80 facets, the King with 86 facets, the Magna cut with 102 facets, and the Royal cut with 144 or 154 facets. Traditionally, each of the brilliant cut facets has its own name (Fig. 8.2).

The quality of diamond 'cuts' is assessed using a number of classifications. The most important stereometric parameters of brilliants are listed in Table 8.1. In keeping with the recommendations of the Diamond High Council (Hoge Raad voor Diamant) in Antwerp, the major European institution issuing quality certificates, the numerical values of these parameters should fall within the ranges specified in the table.

The working of rough diamonds to obtain brilliants is a multi-stage process. Present day computer-aided technology makes it possible to predict the effects of light dispersion in ideal brilliants of known shapes and sizes. In actual practice, however, in spite of the computerization and automation, diamond cutting and polishing — the aim of which is to

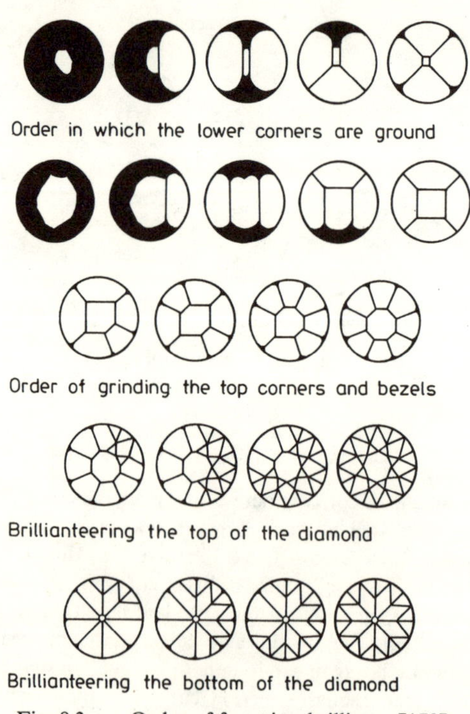

Fig. 8.3 — Order of facetting brilliants [170].

Table 8.1 — Stereometric parameters analysed when evaluating a brilliant cut diamond (Diamond High Council, Antwerp)

	Unusual	Good	Very good	Good	Unusual
Crown height ($\%h_c$)	up to 8,5	9 to 10.5	11 to 15	15,5 to 17	17.5 and up
Table diameter ($\%\ \Phi_t$)	up to 52	53 to 55	56 to 66	67 to 70	71 and up
Crown angle (β)	up to 26.5°	27° to 30.5°	31° to 37°	37.5° to 40°	40.5° and up
Pavilion depth ($\%h_p$) or	up to 39	39.5 to 40.5	41 to 45	45.5 to 46.5	47 and up
Pavilion angle (α)	up to 38°	38.5° to 39°	39.5° to 42°	42.5° to 43°	43.5° and up
Girdle thickness		very thin	thin & med.	thick	very thick
Culet			up to 1.9 %	2% to 3.9%	4% and up

Additional condition for the diamond in order to be considered good or very good. The reflection of the girdle should not be visible through the table-facet by perpendicular observation.

Table 8.2 — The largest known rough diamond crystals [27]*

Carats	Name	Place	Cut into
3106.0	Cullinan	South Africa	Cullinans I–IX, 96 others
995.2	Excelsior	South Africa	21 gems (largest 69.8 ct)
968.9	Star of Sierra Leone	Sierra Leone	Star of Sierra Leone (238.5 ct), 16 others
793.0	Great Mogul	India	Great Mogul (280 ct)
770.0	Woyie River	Sierra Leone	30 gems (largest 31.35 ct)
726.6	Vargas	Brazil	Vargas (48.26 ct), 22 others
726.0	Jonker	South Africa	Jonker (125.65 ct), 11 others
650.25	Reitz	South Africa	Jubilee (245.35 ct)
616.0	Dutoitspan	South Africa	—
609.25	Baumgold Rough	South Africa	14 gems
601.25	Lesotho	Lesotho	17 gems (largest 70 ct)
600.0	Goyaz	Brazil	80-carat gem from one fragment
511.25	Venter	South Africa	32 gems (largest 18 ct)
503.0	Kimberley Rough	South Africa	—
469.0	Victoria 1884	South Africa	Victoria 1884 (185 ct)
455.0	Darcy Vargas	Brazil	—
440.0	Nizam	India	Nizam (277 ct)
434.0	Light of Peace	West Africa	—
428.5	Victoria 1880	South Africa	Victoria 1880 (228.5 ct)
428.5	De Beers	South Africa	De Beers (234.5 ct)
426.5	Ice Queen	South Africa	—
416.25	Berglen	South Africa	—
412.5	Broderick	South Africa	—
410.0	Pitt	India	Regent (140.5 ct)
409.0	Presidente Dutra	Brazil	46 gems (largest 9.06 ct)

* Natural polycrystalline diamonds are not included in the above table — a 3167 ct carbonado diamond was found in Brasil in 1905.

highlight the most advantageous optical properties of a particular crystal — continue to be tantamount to creating a highly specific work of art, and the process of converting a rough stone into a brilliant critically depends on the craftsman's skill, expertise and experience. Diamond working is preceded by careful visual examination of the stone using a loupe, to determine the best way to convert it into a brilliant optimal in respect of aesthetic properties and market value. Following that, any crystal defects are removed by sawing or cleaving, to fit the projected spatial composition. Depending on the desired shape and on the degree of stone perfection, it is then subjected to further sawing, cleaving or bruting, following which it is ground and polished on a diamond-charged cast iron disc called a scaife. The sequence in which the different facets of a brilliant are formed is shown in Fig. 8.3. The loss of rough stone mass will depend on the original shape and state of the stone and on the final cut shape and size aimed at, which may amount to over 50% by weight. It goes without saying that the 'waste' material is carefully collected for other uses. The largest known rough stones and the most famous brilliants are listed in Tables 8.2 and 8.3. The largest rough stone, the Cullinan (pieces of which adorn the British regalia), weighed 3106 ct (= 611.2 g). It took many years to decide how best to convert it into several brilliants. Eventually it was cut into 9 large and 96 small stones.

Table 8.3 — The largest polished diamonds [27]

Carats	Name	Colour	Shape
530.20	Cullinan I	White	Pear
312.40	Cullinan II	White	Cushion
280.00	Great Mogul	White	Rose-cut
277.00	Nizam	White	Dome
245.35	Jubilee	White	Cushion
238.5	Star of Sierra Leone	—	—
234.50	De Beers	Yellow	—
228.50	Victoria 1880	Yellow	Brilliant
205.00	Red Cross	Yellow	Square
199.60	Orloff	White	Rose-cut
185.00	Darya-i-Nur (Iran)	Pink	Table-cut
185.00	Victoria 1884	White	Oval
183.00	Moon	Yellow	Brilliant
152.16	Iranian Yellow A	Yellow	Cushion
150.00	Darya-i-Nur (Dacca)	White	Cushion
140.50	Regent	White	Cushion
137.27	Florentine	Yellow	Double rose
136.50	Queen of Holland	Blue	Cushion
135.45	Iranian Yellow B	Yellow	Cushion
128.80	Star of the South	White	Oval
128.51	Tiffany	Canary	Cushion
128.25	Niarchos	White	Pear
127.00	Portuguese	White	Emerald
125.65	Jonker	White	Emerald

For jewellery, brilliants are usually set in metal 'findings' although some diamonds are in existence as individual works of art, without any setting. The setting of a brilliant and its composition with other stones are to make prominent its beauty. For that reason, the choice of material for the setting and its sculpture, as well as the spatial arrangement of the brilliant(s) and other stones, are of the utmost importance in that they decide the aesthetic value of the whole product. Obviously enough, all this depends on the artist's skill and aesthetic taste. The colour of the setting and of any accompanying stones may produce an apparent change in the colour of the brilliant, which could adversely affect the jewellery piece's market value. Since a gold or brown colour makes diamond look yellowish, the most common setting metal for brilliants is white gold or platinum, and a suitable stone to use with diamonds in the same piece of jewellery is e.g. blue sapphire [130].

9

Economic aspects of diamond use and future prospects

The technological progress of the past decades is closely connected with the steadily growing quantity of diamond employed and continuous progress in the application technologies. One can actually postulate the existence of a certain kind of proportionality between the amount of diamond used by a given economy and the quantity of material goods produced by it [14]. For some time now industrial diamonds have been one of the most essential raw materials in the American, Canadian, Western European, Japanese and Soviet economies. In practice, all of the synthetic and nearly all the natural diamond produced is used up by the economically and industrially most advanced nations [175].

The curve illustrating the world's diamond consumption over the years (Fig. 9.1) is monodirectional: it continues to climb up from one year to the next. At the same time, one can see in it distinct fluctuations coinciding with periods of international tension and conflict. As has already been pointed out, diamond is above all a tool material. It has been recognized for centuries that those who have better tools work more efficiently and, in this way, become stronger and more affluent than those who employ less efficient tools. Just as one talks of the ancient ages of stone, bronze and iron—the materials that made the tools basic to man in those times—our times are sometimes called the age of diamond and other superhard materials.

The steady growth in world diamond consumption results from cost-benefit analysis and from the growing standard of living enjoyed especially in industrially developed nations. Single diamond products

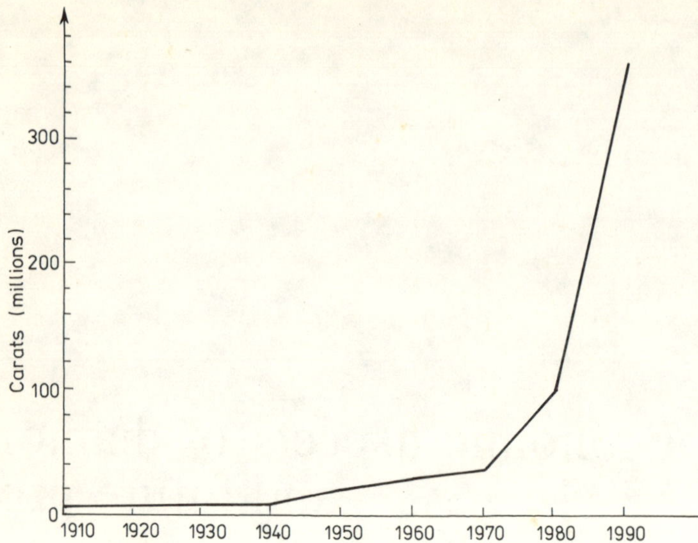

Fig. 9.1 — The rise in consumption of industrial diamonds in the twentieth century 1910–1990 (Daniel (Ed.) — Martin, 1990) [50].

for industrial applications are very costly in terms of initial purchase price when compared to similar tools made from other materials. But the benefits related to their use are considerable—they make it possible to produce more per unit time, and to reduce labour costs. Examples illustrating the correctness of this statement abound in virtually every issue of such specialist publications as *Industrial Diamond Review*, *Industrie Diamanten Rundschau*, *Indiaqua*, or *Sverkhtverdye Materialy*. In his recent book, Ginzburg [98] has analysed various aspects of diamond tool use in Soviet industry over the past several years: it turns out that in the vast majority of cases, conventional tools do not even begin to compare with diamond and other superhard material tools in terms of efficiency of use and the speed and accuracy of operation. The overall cost of converting a unit volume of material into finished products with the use of diamond tools is from several times to several hundred times lower than with conventional tools. The introduction of diamond tools into the Soviet economy has reduced production costs by millions and millions of roubles, while the volume of production has increased. To coordinate the introduction of diamond tools, a number of large, specialized R and D centres have been established. In Western economies, similar effects result from the work above all of the De Beers Industrial Diamond Division. What used to be a mysterious and very rare gemstone has now become a major industrial material in everyday use.

One important factor responsible for the wide industrial use of diamond today is the continuous, world-wide expansion in the use of hard, tough materials such as sintered carbides, stone, concrete, ceramics, glass, etc. For many of the commonly employed hard materials, diamond (in various forms) is virtually the only tool material capable of machining them efficiently. On the other hand, we are living in an age of automation and mass production, circumstances which together enforce increased tolerance requirements. Not infrequently, diamond-based tools (e.g. diamond roller dressers) are the only tools guaranteeing the required repeatability of mass-produced items, the per-item costs being relatively low. Diamond tools are often introduced also to cut the time it takes to do a job, especially if it is done under particularly difficult conditions, such as those prevailing in construction site work or off-shore drilling, or simply to improve working conditions and make a job 'less of a drag'. These tendencies are here to stay, and their importance is likely to grow. In industrially developed countries, human labour is becoming increasingly more costly, and the relative cost of diamond machining thus decreases (Fig. 9.2), making it increasingly more economical.

The growth of the diamond industry involves much more than identification of areas where the mineral can be economically employed, or the development of new tool designs and optimizing the parameters of their useful application. The current growth would not have been possible without the creation of a raw-material base. Strictly speaking,

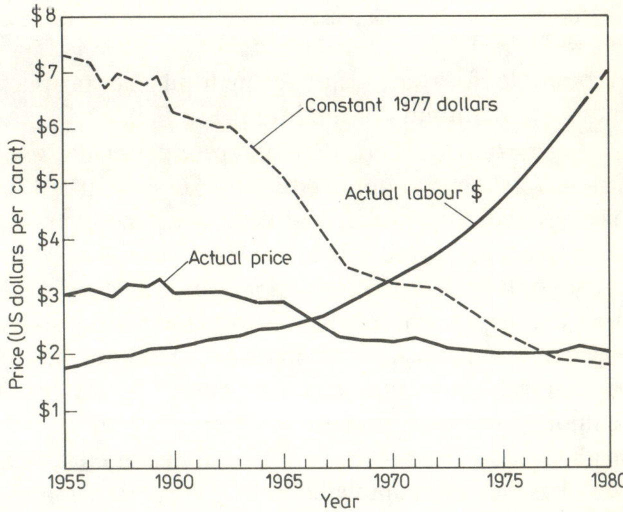

Fig. 9.2 — Diamond abrasive price index, 1955–1980. Also shown is the increase in labour costs (average US hourly earnings including overtime [75].

one can talk of a dynamic growth of the industry only since the development—in the 1950's — of commercially viable methods of synthesis of diamond from graphite. To be sure, natural diamond crystals are still widely used in industry and they have yet to be fully replaced in some optical and electronic industry applications. Still, an increasingly growing proportion of overall diamond consumption is 'man-made'. This tendency too is likely to deepen, since in many cases PCD tool blanks have turned out to be more than adequate substitutes for natural crystals. At the same time, the mass production of synthetic diamonds makes it possible for many countries to become independent of natural diamond sources, confined as these are to a few known places on the Earth. Moreover, while diamond mining runs against natural barriers, the volume of synthetic production can be more readily increased simply by installing the requisite equipment. Thus the relative cost of synthetic diamond—as compared with other major raw materials — has decreased in real terms over the years (Fig. 9.3). This too is

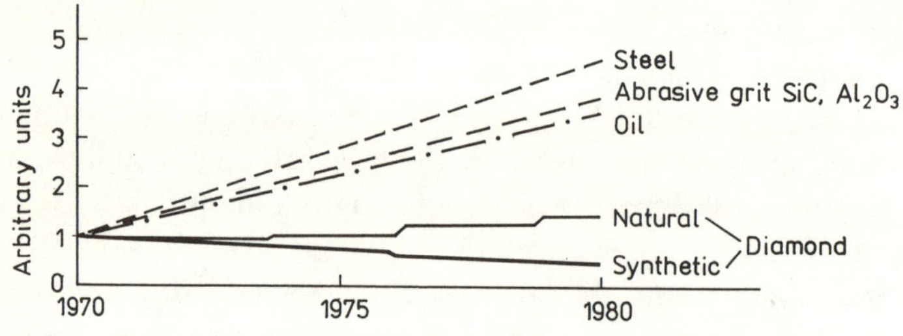

Fig. 9.3 — Comparison between the cost of diamond abrasive and other variables over the decade 1970–1980 (UK statistics) [102].

responsible for the steady growth in the proportion of synthetic industrial diamond consumption (Fig. 9.4).

Progress in synthetic diamond production consists in increasing the volume of diamond obtained from a single synthesis cycle, maintaining or improving its quality, and decreasing the overall production costs. Since fabrication relies heavily on the use of sophisticated ultra-high pressure, high temperature apparatus, any new developments in this area are dependent on new developments in materials engineering, especially with regard to materials resistant to high thermal and mechanical loads. One can thus expect further progress in dynamic diamond synthesis, and in the fabrication of diamond films by the methods described in Section 2.2.4. Laboratory testing and the experience that has been accumulated over the years of large-scale industrial diamond use point to a close interdependence between diamond quality and the economic and technological results deriving from its use. In each case, the work-piece material involved and the machining conditions impose specific requirements upon the mechanical and thermal properties of the diamond. Consequently, quite a wide range of diamond types and PCD blanks has been developed, each appropriate to a specific set of application requirements. Again, this tendency to increasing specialization is likely to grow as should the range of applications of other hard or superhard materials which, in specific cases, may be more economical to use. An example is the use of cubic boron nitride (CBN) for working ferrous materials.

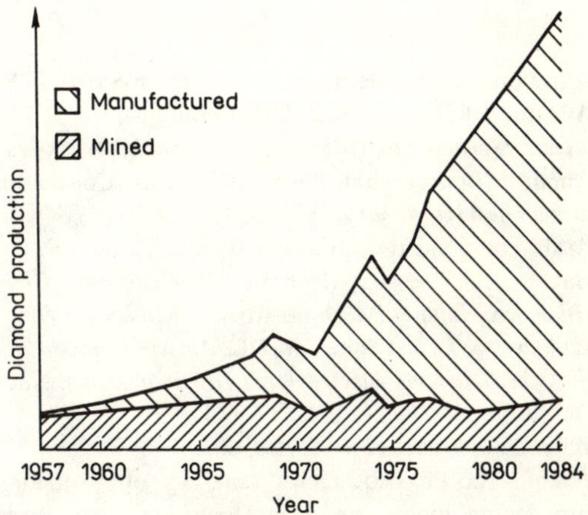

Fig. 9.4 — Relative quantities of natural and synthetic diamond abrasive [145].

Apart from the use of diamonds in hard machining operations, a sizeable proportion — especially of large, high quality monocrystals — is employed in various kinds of research and measuring and control instruments. As the volume and intensity of scientific work on man's immediate and more distant environment (outer space) continues to increase, so will — no doubt — the demand for diamonds.

Industrial uses aside, the most beautiful diamond crystals will certainly continue to be appreciated and sought for as the most highly valued gemstones.

References

[1] Ademant GmbH, Germany, Brochure on SYNDITE cutting tool blanks.
[2] Advanced Resins Limited, UK, Catalogue.
[3] *Almaz: Spravochnik* (*Diamond: Guidebook*), Naukova Dumka, Kiev, 1981.
[4] Almaznie Bezkonechnie Shlifovalnie Lenty (Continuous Diamond Grinding Belts), *Sinteticheskiye Almazy*, 1973, **3**, 93.
[5] Asahi Diamond Industrial Co., Japan, Catalogue.
[6] Babaev, S. G. *et al.*, *Almaznoe Khoningovanie Glubokikh i Tochnykh Otverstii* (*Diamond Honing*), Mashinostroenie, Moskva, 1978.
[7] Baikalov, A. K. and Sukennik, I. L., *Almazy Pravyshchii Instrument na Galvanicheskoi Svyazke* (*Diamond Electroplated Dressing Tools*), Naukova Dumka, Kiev, 1976.
[8] Bakoń, A., *Mechanik*, 1979, **12**, 661.
[9] Bakoń, A., *Mechanik*, 1983, **10**, 603.
[10] Bakoń, A., Physicochemical analysis of synthetic diamonds as a basis for broadening application areas, Doctorate thesis, Wrocław Polytechnic, Wrocław, 1980.
[11] Bakoń, A., *Sverkh. Materialy*, 1983, **6**, 20.
[12] Bakoń, A., *Technika Poszukiwań Geologicznych*, 1987, **26**, 6, 16.
[13] Bakoń, A., Marciniak, A. and Plaskota, T., *Technika Poszukiwań Geologicznych*, 1986, **25**, 2, 1.
[14] Bakul, V. N., *Sinteticheskiye Almazy v Mashinostroenii* (*Synthetic Diamonds in Engineering*), Naukova Dumka, Kiev, 1976.
[15] Bakul, V. N. and Andreev, V. D., *Sinteticheskie Almazy*, 1975, **5**, 3.
[16] Balfour, I., *Famous Diamonds*, De Beers Consolidated Mines Ltd.
[17] Belling, N. and Bialy, L., "The Friatester—10 years later", *Diamond Information Series*, De Beers Industrial Diamond Division, UK.
[18] Belyankina, A. V., *Sinteticheskie Almazy*, 1971, **3**, 22; *ibid.*, 1972, **4**, 20.
[19] Belyankina, A. V. *et al.*, *Sinteticheskie Almazy*, 1975, **7**, 5.
[20] Bergman, O. R. *et al.*, *Metallography*, 1982, **15**, 121.
[21] Berman, R., *Physical Properties of Diamond*, Clarendon Press, Oxford, 1965.
[22] Bogatyreva, G. P., *Sinteticheskie Almazy*, 1977, **8**, 14.
[23] Bokii, G. B. *et al.*, *Prirodnye i Sinteticheskie Almazy* (*Natural and Synthetic Diamonds*), Nauka, Moskva, 1986.
[24] Bowen, G. S. and Seal, M., *Ind. Diam. Rev.*, 1975, **35**, 165.
[25] Bridgman, P. W., *J. Chem. Phys.*, 1947, **15**, 92.
[26] Brugger R., *Nickel Plating*, Robert Draper Ltd., Teddington, 1970.
[27] Bruton, E., *Diamonds*, Chilton Book Company, Radmor, Pennsylvania, 1970.
[28] Bullen, G. J., *Ind. Diam. Rev.*, 1982, **42**, 7.
[29] Bullen, G. J., *Ind. Diam. Rev.*, 1982, **42**, 274.

[30] Bullen, G. J., "The temperature of grinding wheel and work-piece", *Diamond Information Series*, De Beers Industrial Diamond Division, UK.
[31] Bumpcutting and Concrete Grooving, De Beers Industrial Diamond Division, UK.
[32] Bundy, F. P., *Chemistry and Physics of Carbon*, 1973, **10**, 213.
[33] Burchanov, G. S. and Yefimova, Y. V., *Tugoplavkiye Metally i Splavy* (*High-fusible Materials and Alloys*), Metallurgiya, Moskva, 1986.
[34] Busch, D. M., "Machine vibrations and their effect on the diamond wheel", *Diamond Information Series*, De Beers Industrial Diamond Division, UK.
[35] Busch, D. M. and Hill, B. S., "Drilling and sawing concrete with diamond tools", *ibid.*
[36] Busch, D. M. and Walker, R. D., "The sawing of concrete and reinforced concrete with diamond saw blades", *ibid.*
[37] Büttner, A., *Ind. Diam. Rev.*, 1980, **40**, 332.
[38] Carrison, L. C., "Drawing non-ferrous and ferrous wire with polycrystalline diamond dies", *Proc. 52nd Annual Convention of Wire Association International, Inc.*, USA.
[39] Ceramics, De Beers Industrial Diamond Division, UK.
[40] Champion, F. C., *Electronic Properties of Diamond*, Butterworth, London, 1963.
[41] Cheremskoi, P. G. *et al.*, *Sinteticheskie Almazy*, 1971, **5**, 33.
[42] Chistyakov, J. V. *et al.*, *Sverkh. Materialy*, 1982, **1**, 35.
[43] Chistyakov, P. S., *Otdielochno-Abrazivnye Metody Obrabotki* (*Abrasive Methods of Machining*), Vysshaya Shkola, Minsk, 1983.
[44] Christensen Diamond Products GmbH, Germany, Catalogue.
[45] Christyakov, E. M. *et al.*, *Instrument iz Metallizirovanykh Sverkhtverdykh Materialov* (*Tools with Coated Superhard Abrasives*), Naukova Dumka, Kiew, 1982.
[46] Civil Engineering, Construction, De Beers Industrial Diamond Division, UK.
[47] Civil Engineering, Roads and Runways, *ibid.*
[48] Collins, A. T., *Ind. Diam. Rev.*, 1974, **34**, 131.
[49] COMPAX, General Electric Co., USA.
[50] Daniel, P. (Ed.) and Martin, J., *Ind. Diam. Rev.*, 1990, **50**, 291.
[51] Davies, G., *Diamonds*, Adam Hilger Ltd., Bristol–Boston, 1984.
[52] Davis, C. E., "A study of the influences on flat lapping with diamond micron abrasives", *Diamond Information Series*, De Beers Industrial Diamond Division, UK.
[53] Dawson, J. B., *Kimberlites and Their Xenoliths*, Springer-Verlag, Berlin, 1980.
[54] De Beers Natural Diamond Abrasives, De Beers Industrial Diamond Division, UK.
[55] De Beers PCD Cutting Tool Blanks, *ibid.*
[56] De Beers PCD Wire Drawing Die Blanks, *ibid.*
[57] De Beers Synthetic Ultrahard Abrasives, *ibid.*
[58] De Carli, D. S., *Science*, 1961, 133.
[59] Decroly, J. C., "Stone extraction from the quarry using diamond wire", *Diamond Information Series*, De Beers Industrial Diamond Division, UK.
[60] Deriagin, B. V. *et al.*, *Kristallizatsiya Almaza* (*Crystallization of Diamond*), Nauka, Moskva, 1984.
[61] Deriagin, B. V. *et. al.*, *Rost Almaza i Grafita iz Gazovoi Fazy* (*Growth of Diamond and Graphite in Gas Phase*), Nauka, Moskva, 1977.
[62] Diagrit Diamond Tools Ltd., UK, Catalogue.
[63] Diamant Boart SA, Belgium, Catalogue.
[64] Diamant-Schleifscheiben Technologie, De Beers Industrie-Diamanten GmbH, W. Germany.

[65] Diamantwerkzeuge im Strassenbau, *ibid.*
[66] Diamond Coated Abrasives, 3M Company, USA, Catalogue.
[67] Diamond Research 1964, De Beers Industrial Diamond Division, UK.
[68] Diamond Research 1970, *ibid.*
[69] Diamond Research 1974, *ibid.*
[70] Dischler, B. and Brandt, G., *Ind. Diam. Rev.*, 1985, **45**, 131.
[71] Dressing and Truing, De Beers Industrial Diamond Division, UK.
[72] D. Drukker and Zn. NV, Holland, Catalogue.
[73] E. I. Du Pont de Nemours Co., USA, Catalogue.
[74] Dyer, H. B., "An appraisal of diamond as a steel grinding abrasive", *Diamond Information Series*, De Beers Industrial Diamond Division, UK.
[75] Dyer, H. B., *Ind. Diam. Rev.*, 1980, **40**, 401.
[76] Eder, K. G., *Ind. Diam. Rev.*, 1983, **43**, 200.
[77] Elwell, D., *Man-Made Gemstones*, Ellis Horwood Publishers, Chichester, UK, 1979.
[78] Fedoseev, D. V. and Uspenskaya, K. S., *Sinteticheskie Almazy*, 1977, **4**, 18.
[79] FEPA Standards for Diamond Saws, FEPA, France.
[80] Field, J. E., *The Properties of Diamond*, Academic Press, Oxford, 1979.
[81] Finnigan, G., "Machining stone with diamond tools", *Diamond Information Series*, De Beers Industrial Diamond Division, UK.
[82] Fishlock, D., "Electrolytic grinding scores five ways", Diagrit (UK) publication DDT, No. R2.
[83] *Fizycheskaya Khimiya Kondensirovannykh Faz, Sverkhtverdykh Materialov i ikh Granits Razdela* (*Physical Chemistry of Condensed Phase, Superhard Materials and Their Surfaces*), Naukova Dumka, Kiev, 1975.
[84] Fontanella, J. *et al.*, *Appl. Opt.*, 1977, **16**, N11, 2949.
[85] FORMSET, General Electric Co., USA.
[86] Fourth DWMI International Technical Symposium *Diamond and CBN Abrasives*, Proceedings, The O'Hare Inn, Des Plaines, Illinois, USA, 1978.
[87] Frantsevich I. N., *Sverkhtverdye Materialy* (*Superhard Materials*), Naukova Dumka, Kiev, 1980.
[88] Freund, D. G., "Optimizing the efficiency of thin wall core drilling in concrete", General Electric Co., USA.
[89] Fritsch, Dr., W. Germany, Catalogue.
[90] Galitski, V. N. *et al.*, *Sinteticheskie Almazy*, 1976, **2**, 31.
[91] General Electric Co., USA, Catalogues.
[92] GEOSET, General Electric Co., USA.
[93] Gerhäuser, W., *Ind. Diam. Rev.*, 1980, **40**, 321.
[94] Grinzburg, B. I., *Ekonomika i Organizatsiya Primienieniya Sverkhtverdykh Materialov* (*Economics and Organization of Superhard Materials Application*), Naukova Dumka, Kiev 1983.
[95] Glass, De Beers Industrial Diamond Division, UK.
[96] GOST 9206-80 (USSR State Standard).
[97] Götz, F., *Diamanten und Diamantwerkzeuge zum Abrichten von Schleifkörpern*, VDI-Verlag, Düsseldorf, 1968.
[98] "Grinding with diamond abrasives", *Diamond Information Series*, De Beers Industrial Diamond Division.
[99] Grodzinski, P., *Diamond Technology*, NAG Press, London, 1963.
[100] Hanusch, W. and Bergmann, G., *Ind. Diam. Rev.*, 1974, **34**, 126.
[101] Haś, Z. *et al.*, Collection of articles and data on *The Growth of Diamond Films under Metastable Conditions*, Łódz Polytechnic, Poland.

[102] Herbert, S., *Ind. Diam. Rev.*, 1984, **44**, 24.

[103] Herbert, S., *Ind. Diam. Rev.*, 1985, **45**, 251.

[104] Honing, De Beers Industrial Diamond Division, UK.

[105] Hughes, F., "Grinding tungsten carbide dry on low-powered machines", *Diamond Information Series*, De Beers Industrial Diamond Division, UK.

[106] Hughes, F., "Interrupted cut grinding of tungsten carbide with diamond abrasive", *ibid.*

[107] *Identifying Man-Made Gems*, NAG Press, London, 1983.

[108] *Ind. Diam. Rev.*, 1976, No. 2.

[109] *Ind. Diam. Rev.*, 1980, **40**, 405.

[110] *Ind. Diam. Rev.*, 1983, No. 5.

[111] Jeynes, C., *Ind. Diam. Rev.*, 1978, **38**, 14.

[112] Juchem, H. O., *Ind. Diam. Rev.*, 1984, **44**, 209.

[113] Kolomiets, V. V. and Polupan, B. J., *Almaznye Praviashchie Roliki pri Vreznom Shlifovanii Detalei Mashin* (*Diamond Truing and Dressing Tools for Deep Grinding*), Naukova Dumka, Kiev, 1983.

[114] Koyama Y. and Enoki, H., *Ind. Diam. Rev.*, 1982, **42**, 77.

[115] Krasnitsa, W. N., *Melkie Almazy* (*Small Diamonds*), Naukova Dumka, Kiev, 1985.

[116] Kremnia, Z. J., *Skorostnaya Almaznaya Obrabotka Detalei iz Technicheskoi Keramiki* (*Fast Diamond Machining of Technical Ceramics*). Mashinostroenie, Leningrad, 1984.

[117] Krumrei, E. W., "Woodworking applications and performance of COMPAX blank tools", General Electric Co., USA.

[118] Lapping, De Beers Industrial Diamond Division, UK.

[119] Larane, A., *Ind. Diam. Rev.*, 1982, **42**, 336.

[120] Leonov, B. N. *et al.*, *Almazy Prilenskoi Oblasti* (*Diamonds from Prilenska Oblast, USSR*), Nauka, Moskva, 1967.

[121] Liander, H., *Ind. Diam. Rev.*, 1980, **40**, 412.

[122] Linari-Linholm, A. A., *Occurrence, Mining and Recovery of Diamonds*, De Beers Consolidated Mines Ltd., UK.

[123] Lindenbeck, D. A., *Ind. Diam. Rev.*, 1974, **34**, 84.

[124] Loladze, T. N. and Bokuchava, G. W., *Iznos Almazov i Almaznykh Krugov* (*Wear of Diamonds and Diamond Tools*), Mashinostroenie, Moskva, 1967.

[125] Lysanov, V. S., *Elbor v Mashinostroenii* (*CBN in Industry*), Mashinostroenie, Leningrad, 1978.

[126] Macro-Matrix Powders, Kennametal Inc., Canada.

[127] Malogolovets, V. G. and Gatilova, J. G., *Sinteticheskie Almazy*, 1970, **4**, 29.

[128] Maramzin, A. V. *et al.*, *Tekhnicheskie Sredstva dla Almaznogo Bureniya* (*Technology of Diamond Drilling*), Nedra, Leningrad, 1982.

[129] Maślankiewicz, K., *Kamienie szlachetne* (*Gems*), Wyd. Geologiczne, Warszawa, 1982.

[130] Maślankiewicz, K. and Szymański, A., *Mineralogia stosowana* (*Applied Mineralogy*), Wyd. Geologiczne, Warszawa 1976.

[131] Matsumoto, S. *et al.*, *J. of Materials Science*, 1982, **17**, 3106.

[132] McKee, R. L., *Machining with Abrasives*, Van Nostrand Reinhold Co., New York, 1982.

[133] Melzer, H., *Ind. Diam. Rev.*, 1986, **46**, 155.

[134] Metzger, J. L., *Ind. Diam. Rev.*, 1982, **42**, 45.

[135] Metzger, J. L., *Superabrasive Grinding*, Butterworths, London, 1986.

[136] Murowski, V. A. *et al.*, *Sinteticheskie Almazy*, 1971, **5**, 31.

[137] Najdich, Y. W., *Paika i Metallizatsiya Sverkhtverdykh Instrumentalnykh Materialov* (*Brazing and Metallization of Superhard Materials*), Naukova Dumka, Kiev, 1977.

[138] "New flexible tool concept promises increased scope for diamond utilization", *Ind. Diam. Rev.*, 1974, **34**, 358.

[139] Niesmielov, A. F., *Narzędzia diamentowe w przemyśle* (*Diamond Tools in Industry*), WNT, Warszawa, 1967.

[140] Norling. G., "Rock drilling with diamond", *Diamond Information Series*, De Beers Industrial Diamond Division, UK.

[141] Norton-Christensen, Inc.., USA Catalogues.

[142] Norton Company, USA, Catalogues.

[143] Novikov, N. W. (Ed.), *Sinteticheskie Sverkhtverdye Materialy* (*Synthetic Superhard Materials*), Naukova Dumka, Kiev, 1986.

[144] Oates, P. D., "Evaluation of cutting fluids for use with ABN abrasives", De Beers Industrial Diamond Division, UK.

[145] O'Brien, P. A., "Concrete sawing and drilling with diamond tools", Abstracts "The world of concrete 83", Las Vegas 1983, USA.

[146] O'Carroll. B. G., "ANSI/FEPA screening brings important benefits to the diamond abrasives industry", *Diamond Information Series*, De Beers Industrial Diamond Division, UK.

[147] Orlov. Y. L., *Morfologiya Almaza* (*Morphology of Diamond*), Izd. AN SSSR, Moskva, 1973.

[148] Orlov, Y. L., *The Mineralogy of the Diamond*, J. Wiley and Sons, London–New York.

[149] Pachalin, I. A., *Almaznoe Kontaktno-erozionnoe Shlifovanie* (*Diamond Electromachining*), Mashinostroenie, Leningrad, 1985.

[150] Pagel-Theisen V., *Diamond Grading ABC*, Rubin and Son, New York.

[151] Patent: USSR 986763.

[152] Peter, H., *Ind. Diam. Rev.*, 1983, **43**, 320.

[153] Petrdlik, M., *Vysokotlaka Technika pri Synteze Materialu* (*High Pressure Technology in Material Synthesis*), PPM-VUPM, 1980, **1**, 21.

[154] Pinzari, M., *Ind. Diam. Rev.*, 1980, **40**, 58.

[155] Pipkin, N. J., *Ind. Diam. Rev.*, 1983, **43**, 231.

[156] Plastics and Rubber, De Beers Industrial Diamond Division, UK.

[157] Popov, S. A., *et al.*, *Almazno-abrazivnaya Obrabotka Metallov i Tverdykh Splavov* (*Diamond Machining of Metals and Hard Alloys*), Mashinostroenie, Moskva, 1977.

[158] *Poradnik galwanotechnika* (*Electroplating Guidebook*), WNT, Warszawa, 1984.

[159] *Poroshki, Instrument i Pasty iz Sinteticheskikh Almazov* (*Synthetic Diamond Powders, Tools and Pastes*), ISM-USSR, Catalogue. Naukova Dumka, Kiev, 1981.

[160] *Proceedings of the Fifth Conference on Carbon*, Pergamon Press, Oxford, 1962.

[161] *Proceedings, De Beers Düsseldorf Conference '79*, De Beers Industrial Diamond Division, UK.

[162] Products for the Stone Industry, Norton Company, USA.

[163] Prokopchuk, B. I., *Almaznye Rossypi i Metodika ikh Prognozirovaniya i Poiskov* (*Diamond Deposits and Methods of Their Output*), Nedra, Moskva, 1979.

[164] Prudnikov, J. L. and Fadeev, V. F., *Sinteticheskie Almazy*, 1975, **5**, 20.

[165] *Radiatsionnie Effekty v Tverdykh Telakh* (*Radiation Effects in Solids*), Naukova Dumka, Kiev, 1977.

[166] Read, P. G., *Ind. Diam. Rev.*, 1980, **5**, 167.

[167] Research and Development 1970, General Electric Co., USA.

[168] Reznikov, A. N., *Abrazivnaya i Almaznaya Obrabotka Materialov* (*Abrasive and Diamond Machining*), Mashinostroenie, Moskva, 1977.

[169] Rock Drilling, De Beers Industrial Diamond Division, UK.

[170] Rubin and Zoon, Antwerp, Catalogue.

[171] Rudenko A. P. *et al.*, *DAN SSSR*, *Ser. mat. fiz.*, 1965, **163**, No. 5, 1169.

[172] Rumyantsev E. V. and Davydov A. D., *Tekhnologiya Elektrokhimicheskoi Obrabotki Metallov* (*Technology of Electrochemical Machining*), Vysshaya Shkola, Moskva, 1984.

[173] Samsonov, G. V. *et al.*, *Poluchenie i Metody Analiza Nitridov* (*Manufacture and Analysis of Nitrides*), Naukova Dumka, Kiev, 1978.

[174] Sappok, R. and Boehm, H. P., *Carbon*, 1968, 283; *ibid.*, 1968, 573.

[175] Schmidt, W. and Malzahn, W., *Industriemineral Diamant*, VEB Deutscher Verlag für Grundstoffindustrie, Leipzig, 1980.

[176] Schneider, W., *Ind. Diam. Rev.*, 1985, **45**, 242.

[177] Schröder, W. and Gaipel, W., *Ind. Diam. Rev.*, 1976, **6**, 200.

[178] Schuman, W., *Gemstones of the World*, NAG Press, London.

[179] Seal, M., *Diamond Heat Sinks for Semiconductor Devices*, D. Drukker and Zn. NV, Holland.

[180] Seal, M., *Ind. Diam. Rev.*, 1971, **31**, 464.

[181] Seal, M., *Naturdiamant als Werkstoff für Wissenschaftliche und technische Anwendungen*, D. Drukker and Zn. NV, Holland.

[182] Semko, M. F., *Osnovy Almaznovo Shlifovaniya* (*Basis of Diamond Grinding*), Tekhnika, Kiev, 1978.

[183] Shafto, G. R. and Notter, A. T., "Truing and dressing of resin bond peripheral wheels", *Diamond Information Series*, De Beers Industrial Diamond Division.

[184] Shchegolev, V. A. and Ulanova, M. J., *Elastichnye Abrazivnye i Almaznye Instrumenty* (*Flexible Abrasive and Diamond Tools*), Mashinostroenie, Leningrad, 1977.

[185] Silveri, P., *Ind. Diam. Rev.*, 1980, **40**, 464.

[186] J. K. Smit and Sons Ltd., UK, Catalogue.

[187] Smith, N. R., *Industrial Applications of Diamond*, Hutchinson and Co., London, 1965.

[188] Sokołowska, A., *Zeszyty Naukowe Akademii Górniczo-Hutniczej, Ceramika*, 1983, **45**, 906.

[189] Stankoimport, USSR, Catalogue.

[190] Stone, De Beers Industrial Diamond Division, UK.

[191] STRATAPAX, General Electric Co., USA, Catalogue.

[192] *Sverkhtverdye Materialy — Sintez, Svoistva, Primienienie* (*Superhard Materials — Synthesis, Properties, Applications*), Naukova Dumka, Kiev, 1983.

[193] SYNDITE — Polykristalliner Diamant zum Verschleissschutz, De Beers Industrie Diamanten GmbH, W. Germany.

[194] SYNDRILL und SYNDAX-3 — Polykristalliner Diamant zum Gesteinsbohren, *ibid.*

[195] Synthetic Diamond Abrasives, General Electric Co., Catalogues.

[196] Szymański, A. and Szymański, J. M., *Badania twardości minerałów, skał i produktów ich przeróbki* (*Hardness Testing of Minerals and Rocks*), Wyd. Geologiczne, Warszawa, 1976.

[197] Thakur, B. N., *Ind. Diam. Rev.*, 1977, **37**, 91.

[198] The Industrial Diamond Association (IDA) of America, Inc., 1977.

[199] Thiel, N. W., "The wet grinding of tungsten carbide and steel combinations with diamond wheels", *Diamond Information Series*, De Beers Industrial Diamond Division, UK.

[200] Thoreau B., *Ind. Diam. Rev.*, 1984, **44**, 94.

[201] Tolansky, S., *Surface Microtopography*, Longmans, London, 1960.
[202] Tomlinson, P. N. *et al.*, "CDA-M—a new dry grinding abrasive for tungsten carbide", *ibid.*
[203] Tomlinson, P. N. *et al.*, "CDA—the new high performance diamond abrasive for grinding carbide", *Diamond Information Series*, De Beers Industrial Diamond Division, UK.
[204] Tomlinson, P. N. *et al.*, *Ind. Diam. Rev.*, 1985, **45**, 299.
[205] Trancu, T. C., "Diamond wire machine cuts marble quarrying costs", *Diamond Information Series*, De Beers Industrial Diamond Division, UK.
[206] Tuzzeo, J. J., "Correlation of diamond properties and performance in surfacing of ophthalmic glass lenses", General Electric Co., USA.
[207] Tyrolit, Austria, Catalogue.
[208] Urbanek, Joh., Germany, Catalogue.
[209] Vakser, D. B. *et al.*, *Almaznaya Obrabotka Tekhnicheskoi Keramiki* (*Diamond Machining of Technical Ceramics*), Mashinostroenie, Leningrad, 1976.
[210] Vdovykin, G. P., *Almazy v Meteoritach* (*Diamonds in Meteorites*), Nauka, Moskva, 1970.
[211] Vermeulen, L. A., South African Patent Application, No. 74/7725.
[212] Wedlake, R. J., *Ind. Diam. Rev.*, 1977, **37**, 196.
[213] WEMA, Poland, Catalog of Polish Tool Industry.
[214] Wendt, Germany, Catalogue.
[215] Wermusch, G., *Adamas*, Verlag die Wirtschaft, Berlin, 1984.
[216] Wilks, E. M. and Wilks, J., *J. Appl. Phys.*, 1972, 5.
[217] Williams, A. F., *The Genesis of the Diamond*, Ernest Benn, London, 1932.
[218] E. Winter and Sohn, Germany, Catalogue.
[219] E. Winter and Sohn, Germany, Notes on Tool Applications.
[220] Yashecheritsyn, P. I., *Elektroerozionnaya Pravka Almazno-abrazivnykh Instrumentov* (*Electro-Erosion Truing and Dressing*), Nauka i Tekhnika, Moskva, 1981.
[221] Young, B., "The graphitization of diamond during the manufacture of diamond tools", *Diamond Information Series*, De Beers Industrial Diamond Division, UK.
[222] Zakharenko, I. P., *Osnovy Almaznoi Obrabotki Tverdo-splavnogo Instrumenta* (*Diamond Machining of Tools with Hard Alloys*), Naukova Dumka, Kiev, 1981.
[223] Zsolnay, L. M., *Ind. Diam. Rev.*, 1977, **37**, 382.

Index

Trade marks:
* — Soviet diamonds
** — De Beers
*** — General Electric
**** — Cemat'70 — Poland
***** — Winter — Germany
****** — Sumitomo — Japan
******* — Advanced Resins Ltd. — UK

Various kinds of industrial diamond stones (courtesy of Henri Polak Company, New York).
(pp. 241–244)

Various kinds of diamond synthetic abrasives (courtesy of De Beers)
(pp. 245–248)

MDA

EDC

CDA